Chuanren Wu

Conceptual Studies of Multistage Depressed Collectors for Gyrotrons

Karlsruher Forschungsberichte aus dem
Institut für Hochleistungsimpuls- und Mikrowellentechnik

Herausgeber: Prof. Dr.-Ing. John Jelonnek

Band 13

This work has been carried out within the framework of the EUROfusion Consortium and has received funding from the Euratom research and training programme 2014-2018 and 2019-2020 under grant agreement No 633053. The views and opinions expressed herein do not necessarily reflect those of the European Commission.

Conceptual Studies of Multistage Depressed Collectors for Gyrotrons

by
Chuanren Wu

KIT Scientific Publishing

Karlsruher Institut für Technologie
Institut für Hochleistungsimpuls- und Mikrowellentechnik

Conceptual Studies of Multistage Depressed Collectors for Gyrotrons

Zur Erlangung des akademischen Grades eines Doktors der Ingenieurwissenschaften von der KIT-Fakultät für Elektrotechnik und Informationstechnik des Karlsruher Instituts für Technologie (KIT) genehmigte Dissertation

von M.Sc. Chuanren Wu

Tag der mündlichen Prüfung: 27. Februar 2019
Referent: Prof. Dr.-Ing. John Jelonnek
Korreferent: Prof. Dr.-Ing. Heino Henke

Impressum

KIT Scientific Publishing

Karlsruher Institut für Technologie (KIT)
KIT Scientific Publishing
Straße am Forum 2
D-76131 Karlsruhe

KIT Scientific Publishing is a registered trademark
of Karlsruhe Institute of Technology.
Reprint using the book cover is not allowed.

www.ksp.kit.edu

Print on Demand 2019 – Gedruckt auf FSC-zertifiziertem Papier

ISSN 2192-2764
ISBN 978-3-7315-0934-9
DOI 10.5445/KSP/1000094598

Foreword of the Editor

Nuclear fusion is considered as a possible option for generating sustainable, safe, CO_2-free, baseload energy in the future. The magnetically confined fusion requires auxiliary heating, in order to reach and sustain a high plasma temperature. The toroidal plasma current, which is indispensable in a tokamak reactor, can be driven non-inductively by external sources for a continuous operation of the tokamak. The heating and current drive of the plasma can take place at the electron cyclotron resonance with high-power microwaves. In addition, the electron cyclotron resonance heating and current drive (ECRH & CD) is the only known system today, that is able to stabilize a very localized plasma. Based on that, the DEMOnstration fusion power plant (DEMO) will need an ECRH & CD system delivering at least 50 MW of microwave power at frequencies up to 240 GHz. Hence, the final plug-in efficiency of that ECRH & CD system is vital for the fusion power gain. Gyrotrons are the only known microwave sources that are capable of efficiently generating continuous-wave (CW) power at multi-megawatt levels. As the gyrotron is the RF source, its efficiency dominates the total efficiency of the ECRH & CD system. Today, the achievable efficiency of a fusion gyrotron is approximately 50 % as known from ITER. To achieve a sufficient power gain for DEMO, the gyrotron efficiency will have to be increased to significantly higher than 60 %.

Recuperation of the energy of the spent electrons is the method of choice to increase the efficiency of any vacuum tube. Today's gyrotrons employ a single-stage depressed collector. A multistage depressed collector (MDC) would increase the overall efficiency significantly. Today, MDCs are widely used in traveling-wave tubes (TWT). However, the TWT MDC principle does not work for high-frequency gyrotrons, mainly due to the high magnetic field enclosed in the gyrotron hollow electron beam and the spread of the kinetic energy of the electrons after the interaction with the RF wave. Hence, it is non-trivial to find an appropriate MDC concept for gyrotrons. To approach the first gyrotron MDC in the world, a series of innovations are required.

For the first time, this work provides a systematical classification of possible MDC concepts for gyrotrons and includes a comprehensive study on all these concepts. It shows that an MDC, that is using axisymmetric *E*- and *B*-fields, is not appropriate for fusion gyrotrons. In opposite to that, this present work demonstrates different concepts of MDCs with non-axisymmetric *E*- and *B*-fields that will fulfil the requirements of an MDC for DEMO gyrotrons. In this frame, Chuanren Wu presents a systematical investigation on the possibilities to create and to use the E×B drift in gyrotron collectors. The most appropriate MDC type for DEMO gyrotrons and the basic ideas for the mechanical realization of a gyrotron E×B MDC are proposed for the first time.

Zusammenfassung

Die auf magnetischem Einschluss basierende Kernfusion stellt eine vielversprechende Methode zur nachhaltigen, emissionsfreien Energiegewinnung dar. Die Elektron-Zyklotron-Resonanz-Heizung (ECRH) und der entsprechende nicht-induktive Stromtrieb (ECCD) koppeln die Leistung einer hochfrequenten (HF) elektromagnetischen Welle in eine gezielte Position im Fusionsplasma ein, um das Plasma zu heizen und zu stabilisieren. Für das EC-System eines Demonstrationskraftwerks (DEMO) ist eine Ausgangsleistung von mindestens 50 MW geplant. Außerdem kann durch ECCD Plasmastrom getrieben werden. Soll der gesamte DEMO-Plasmastrom durch ECCD getrieben werden, wird eine HF-Leistung von deutlich mehr als 100 MW benötigt. Um einen Netto-Leistungsgewinn des Kraftwerks zu demonstrieren, ist die elektrische Eingangsleistung des Plasmaheizungs- und Stromtriebsystems für die Leistungsbilanz kritisch. Das Gyrotron ist die HF-Quelle für ECRH und ECCD. Der Wirkungsgrad eines Gyrotrons ist daher wesentlich für DEMO. Unter Berücksichtigung der Verluste im HF-Übertragungs-System, sowie in den anderen Komponenten des EC-Systems, sollte die Effizienz eines Gyrotrons deutlich höher als 60 % sein, um einen 60-prozentigen Wirkungsgrad des DEMO-EC-Systems zu erreichen.

Heutzutage werden in Gyrotrons einstufige Kollektoren mit Gegenspannung zur Energierückgewinnung („Depressed Collector") eingesetzt. Ein modernes Gyrotron dieser Art hat einen gesamten Wirkungsgrad von circa 50 %. Um den Wirkungsgrad noch deutlich zu erhöhen, benötigt man Kollektoren mit mehrstufiger Energierückgewinnung (sogenannte MDCs). Entsprechend besitzen diese Kollektoren mehrstufige Steigerungen der Gegenspannung, um mehr kinetische Energien aus den Elektronen zurückzugewinnen. MDCs werden seit vielen Jahren erfolgreich in Wanderfeldröhren eingesetzt. Jedoch ist es bis heute nicht gelungen, einen mehrstufigen Kollektor für Gyrotrons zu testen. Allerdings hat es bereits intensive Bemühungen gegeben, so können einige Entwürfe für MDCs in der Literatur gefunden werden. Die größte Herausforderung für einen mehrstufigen Kollektor besteht darin, dass die Elektronen im Gyrotronkollektor immer noch durch ein starkes Magnetfeld geführt werden. Dies bedeutet, dass der gesamte Hohlstrahl einen hohen magnetischen Fluss umschließt. Die zweite Herausforderung besteht darin, dass die Elektronen nach der Wechselwirkung eine große Streuung der kinetischen Energien besitzen. Daher ist es nicht möglich, das Prinzip des mehrstufigen Kollektors für eine Wanderfeldröhre auf das Gyrotron zu übertragen. Aufgrund des hohen magnetischen Flusses würde dies unmittelbar zu einer unrealistisch großen Kollektorgeometrie führen. Genau dies zeichnet alle bisher bekannten Entwürfe von MDCs für Gyrotrons aus. Infolgedessen sollen neuartige Konzepte untersucht werden, die auf alternativen physikalischen Prinzipien basieren.

Das vorrangige Ziel dieser Arbeit ist die systematische Untersuchung physikalischer Prinzipien für mehrstufige Kollektoren für Gyrotrons. Dabei zeigt sich, dass diese in zwei Klassen unterteilt werden können. Die erste Klasse beinhaltet diejenigen Kollektoren, für die das elektrische Potential und das magnetische Vektorpotential axialsymmetrisch sind. Die zweite Klasse enthält diejenigen Kollektorarten, bei denen die Elektronen anhand der sogenannten $\boldsymbol{E}\times\boldsymbol{B}$ Drift sortiert werden.

Alle aus der Literatur bekannten Entwürfe gehören zur ersten Klasse von MDCs. Wie bereits erwähnt, sind diese Entwürfe ungeeignet für Gyrotrons, die bei hohen Frequenzen, also bei entsprechend hohen magnetischen Feldern betrieben werden. Bezüglich MDCs mit axialsymmetrischen Feldern konzentriert sich diese Arbeit daher auf einen spezifischen nicht-adiabatischen Übergang des Magnetfelds im Kollektor. Die Elektronen werden entsprechend ihrer kinetischen Energie den verschiedenen Kollektorstufen zugeordnet. Dabei ist die Änderung des axialen Magnetfelds mehr nicht-adiabatisch als in allen bisher bekannten Entwürfen. Dies führt dazu, dass der in dieser Arbeit vorgeschlagene axialsymmetrische Kollektorentwurf zum ersten Mal eine realisierbare geometrische Größe haben kann. Allerdings besitzt auch dieses Konzept Einschränkungen. Sie werden in dieser Arbeit mittels realistischer 3D PIC- sowie Laufbahnsimulationen demonstriert. Ein zweistufiger axialsymmetrischer Kollektor wird höchstwahrscheinlich das Ziel eines Gesamtwirkungsgrads von 60 % für das Gyrotron nicht erreichen. Eine Erweiterung zu mehr als zwei Kollektorstufen ist ebenfalls unrealistisch. Der Grund dafür ist die Erhaltung des Impulsmoments, so dass die Elektronenlaufbahnen nach diesem Konzept nicht trennbar sind.

Die zweite Klasse von MDCs, die auf nicht-axialsymmetrischen elektrischen und magnetischen Feldern beruht und das Prinzip der $\boldsymbol{E}\times\boldsymbol{B}$ Drift für die Sortierung der Elektronen nutzt, kann dagegen die Elektronen, deren kinetische Energie zu unterschiedlichen Energieintervallen gehören, eindeutig trennen. Obwohl die $\boldsymbol{E}\times\boldsymbol{B}$ Drift allgemein einfach verständlich ist, ist die Umsetzung dieses Prinzips nicht selbsterklärend. Realistische Möglichkeiten zur Umsetzung in Gyrotrons existieren nach dem Wissen des Autors nicht. In dieser Arbeit werden deshalb vier bisher unbekannte und daher innovative Entwürfe vorgestellt. Die jeweiligen Vor- und Nachteile werden anhand realistischer Simulationen diskutiert. Ein erfolgversprechender Entwurf zeichnet sich dabei nicht nur durch hohen Wirkungsgrad aus, sondern es wird ebenfalls eine genügende Unempfindlichkeit gegenüber Störfeldern bzw. mechanischem Versatz erwartet. Außerdem weist ein derartiger Entwurf die niedrigsten Anforderungen an die notwendigen Kollektorspulen auf und ist ausreichend kompakt gebaut. Eine Idee für die mechanische Implementierung wird zum ersten Mal in dieser Arbeit präsentiert.

Abstract

The Electron Cyclotron Resonance Heating and Current Drive (ECRH and ECCD) system can deposit the Radio-Frequency (RF) heating power at localized positions desired by the nuclear fusion plasma. For the DEMOnstration fusion power plant, an ECRH power of 50 MW is planned. If the DEMO plasma current should be driven by ECCD, an RF power of higher than 100 MW is expected due to the large DEMO plasma volume. In order to demonstrate a net power gain of the DEMO plant, the input power consumed in the heating and current drive is critical. Gyrotron is the RF source of the ECRH and ECCD system. The efficiency of a gyrotron is therefore vital for the DEMO power plant. Taking into account the losses in the RF transmission as well as other parts of the EC system, a gyrotron should have an efficiency of significantly higher than 60 %, in order to achieve an EC system efficiency of 60 % for DEMO.

Nowadays the gyrotrons with single-stage depressed collectors have approximately a total efficiency of 50 %. In order to significantly raise that efficiency, a Multistage Depressed Collector (MDC) should be applied. MDCs have been successfully applied in Traveling-Wave-Tubes (TWTs). However, no MDC for gyrotrons exists in the world. There were a few researches and design proposals of gyrotron MDCs prior to this work, but none of them has been successfully operated. The main challenge of a gyrotron MDC comes from the moderate magnetic field in the gyrotron collector region and the high magnetic flux enclosed in the annular spent electron beam, which has a broad kinetic energy distribution. Adapting the MDC concepts from TWTs requires an additional huge volume in order to scale down the magnetic flux density in the collector of a gyrotron. That is unpractical. The conventional gyrotron MDC proposals in the existing researches will also have the same issue for a high-frequency gyrotron. Hence, new MDC concepts based on other physical principles should be investigated, which will be more applicable in the fusion gyrotrons.

This work presents a systematic investigation on two possible fundamentally different MDC types for gyrotrons: in the first type, the electric potential and the magnetic vector potential are axisymmetric, while the other non-axisymmetric type makes use of the $\boldsymbol{E}\times\boldsymbol{B}$ drift to sort electrons according to their kinetic energy.

The conventional proposals in the previous researches belong to the axisymmetric type. As mentioned before, those design proposals were not practical to be up-scaled for high-frequency (high magnetic flux) gyrotrons. A new variant of axisymmetric MDC will be conceptually designed and investigated in this work. This new concept requires a special non-adiabatic transition of magnetic field to sort the local phases of the electron cyclotron motions. The magnetic field is much more non-adiabatic than in the existing proposals. Therefore, the expected size of such a collector is smaller than the one up-scaled from the existing proposals. The limitations of this new axisymmetric MDC proposal will be demonstrated with simulations. A two-stage axisymmetric collector may not achieve the goal of a 60 % total gyrotron efficiency, while upgrading to more than two stages for a higher efficiency is especially challenging, because there is no separation of electron orbits due to the conservation of canonical angular momentum.

The MDC type based on the $\boldsymbol{E}\times\boldsymbol{B}$ drift has a distinct sorting mechanism, where the electrons belonging to different energy intervals can be totally separated. Therefore, it is considered to be more advantageous than its axisymmetric counterpart. Although the physical principle is straightforward, the conversion of the physical principle to a conceptual collector design is non-trivial and was not explored prior to this work. In the presented work, four innovative designs will be proposed. Their advantages and disadvantages will be discussed based on the systematic investigations. The most promising design is expected to be efficient, insensitive to the operation environment, can be built under a reasonable constraint of size, and has a low requirement on additional electromagnetic coils. A mechanical design proposal of that will be presented at the end of this work.

The finally proposed MDC design is the most promising one among all gyrotron MDC proposals up to now. The principle and concept should be experimentally verified in the next step for the future DEMO gyrotrons.

Contents

Foreword of the Editor . . . . **i**

Zusammenfassung . . . . **iii**

Abstract . . . . **vii**

Abbreviations . . . . **xiii**

Symbols . . . . **xv**

1 Introduction **1**

1.1 Motivation 1

1.2 Gyrotrons for Nuclear Fusion 2

1.2.1 Principle of Nuclear Fusion 2

1.2.2 Fundamentals of Fusion Gyrotrons 5

1.2.3 Theoretical Estimation of the Gyrotron Efficiency 9

1.3 Basic Physical Principles 12

1.3.1 Electron Motion in Electrostatic and Magnetostatic Fields . 12

1.3.2 Emission and Suppression of Secondary Electrons 16

1.4 Technical Principles of Energy Recuperation in Vacuum Tubes . . 21

1.4.1 Basic Principle of Depressed Collectors 21

1.4.2 Multistage Depressed Collectors for Linear Beam Tubes . . 29

1.4.3 State-of-the-Art for Gyrotron Multistage Depressed Collectors 34

1.5 Critical Parameters and Constraints of the Multistage Depressed Collector Designs in this Work 41

1.6 Collector Simulation Techniques and Codes Used in this Work . . 44

2 Proposal for an Advanced Axisymmetric Collector Concept Allowing a High Magnetic Flux within the Electron Beam 49

2.1 Advances to the State-of-the-Art in Brief 49

2.2 New Principle for the Sorting of Electron Energies in an Axisymmetric System 50

2.3 Analysis of the Electron Energy Sorting Model 52

2.3.1 Introduction of the Theoretical Model 52

2.3.2 Maximal Reachable Radius as a Function of Electron Velocity 55

2.3.3 Sorting of Guiding Centers 57

2.3.4 Sorting of Larmor Radii 58

2.3.5 Sorting of Longitudinal Cyclotron Periods 61

2.3.6 Properties of the Non-Adiabatic Transition Learned from this Analysis 61

2.4 Conceptual Collector Design and Basic Examinations 62

2.4.1 An Optimized Axisymmetric Collector Design for a 170 GHz 1 MW Gyrotron 62

2.4.2 Examination of the Effective Potential 65

2.4.3 Examination of the Energy Sorting Quality Using Mono-Energetic Electron Beams 67

2.5 Assessment of the Critical Parameters 68

2.5.1 Achievable Collector Size and Mechanical Complexity . . . 69

2.5.2 Expected Collector Efficiency and Unwanted Back-Stream Current 69

2.5.3 Maximal Expected Power Load Density on the Critical Parts 75

2.5.4 Summary 77

3 Proposals for Non-Axisymmetric Collector Concepts Based on *E*×*B* Drift **79**

3.1 Advances to the State-of-the-Art in Brief 79

3.2 Possibilities to Use the ***E***×***B*** Drift in a Gyrotron Collector 80

3.2.1 Using a Planar Drift 81

3.2.2 Using a Radial Drift 83

3.3 Proposal I: Planar Drift Applied to Sheet Electron Beams 87

3.3.1 Creation of Sheet Electron Beams 87

3.3.2 Conceptual Collector Design and Basic Examinations . . . 89

3.3.3 Optimization of Electromagnetic Coils Based on the Magnetic Flux Coordinates 91

3.3.4 Upgraded Designs of Coils and Electrodes 98

3.3.5 Assessment of the Critical Parameters 100

3.4 Proposal II: Planar Drift Applied to an Elliptic Electron Beam . . 104

3.4.1 Creation of an Elongated Elliptic Electron Beam 104

3.4.2 Estimation of the Minimal Electrode Length 106

3.4.3 Assessment of the Critical Parameters 107

3.5 Proposal III: Planar Drift Applied to Arc Electron Beams 109

3.5.1 Creation of Arc Electron Beams 109

3.5.2 Estimation of the Total Collector Length 111

3.5.3 Conceptual Collector Design and Basic Examinations . . . 114

3.5.4 Assessment of the Critical Parameters 116

3.6 Proposal IV: Radial Drift Applied to the Annular Electron Beam . 121

3.6.1 Creation of a Radial Drift 121

3.6.2 Conceptual Collector Design and Basic Examinations . . . 123

3.6.3 Assessment of the Critical Parameters 127

3.7 Upgrades for Proposal IV 135
3.7.1 Upgraded Design Proposals to Suppress the Back-Stream Current of Proposal IV 135
3.7.2 The Final Upgrade for a Reduction of the Collector Size . . 143
3.8 Comparison of the Design Proposals in this Chapter 144
3.9 Proposal for the Mechanical Design of a Two-Stage Depressed Collector 146

4 Conclusion and Outlook **151**

A Theoretical Supplements **157**
A.1 Dynamic Programming Algorithm for the Estimation of the Maximal Collector Efficiency 157
A.2 Evaluation of Magnetic Field and Vector Potential Induced by Electromagnetic Coils 159
A.3 An Ideal Smooth Transition of Magnetic Field 166
A.4 Derivation of the Drift Distance for Proposal IV 168

B Additional Collector Design Proposals **171**
B.1 An Existing $\boldsymbol{E}\times\boldsymbol{B}$-Type of Design Proposal Using Azimuthal Magnetic Field 171
B.2 A Different Kind of Axisymmetric Design Proposal Using Electric Field to Sort Electrons 172
B.3 An Additional $\boldsymbol{E}\times\boldsymbol{B}$-Type of Design Proposal Using Inverted Helical Electrodes 174

Bibliography **177**

List of Figures **191**

Abbreviations

a.u.	arbitrary unit
ASDEX	Axially Symmetric Divertor Experiment
BREP	Boundary REPresentation
BWO	Backward Wave Oscillator
CSG	Constructive Solid Geometry
CW	Continuous Wave
DC	Direct Current
DEMO	DEMOnstration fusion power plant
DP	Dynamic Programming
EC	Electron Cyclotron
ECCD	Electron Cyclotron Current Drive
ECRH	Electron Cyclotron Resonance Heating
FC	Faraday Cup
FEL	Free Electron Laser
FULGOR	Fusion Long Pulse Gyrotron Laboratory
GND	Ground potential
IMT	the quadrature transformation proposed by M. Iri, S. Moriguti and Y. Takasawa
ITER	International Thermonuclear Experimental Reactor

KIT	Karlsruhe Institute of Technology
m.u.	Mesh Unit
MDC	Multistage Depressed Collector
MIG	Magnetron Injection Gun
MOU	Matching Optics Unit
NB, NBI	Neutral Beam (Injection)
NURBS	Non-Uniform Rational B-Splines
PC	Personal Computer
PDF	Probability Density Function
PIC	Particle In Cell
QO	Quasi-Optical
RF	Radio Frequency
SC	Super Conducting
SDC	Single-stage Depressed Collector
SEY	Secondary Electron Yield
TE	Transverse Electric
TEF	Tilted Electric Field
TL	Transmission Line
TM	Transverse Magnetic
TRK	Trajectory Tracking
TWT	Traveling Wave Tube
W7-X	The Wendelstein 7-X stellarator

Symbols

$\boldsymbol{A}$	magnetic vector potential
A_θ	azimuthal component of magnetic vector potential
a	semi-major axis of an ellipse
a_t	short for tan α_t, see α_t
$\boldsymbol{B}$	vector of magnetic flux density
B	magnetic flux density
B_0	initial value of B
B_max	maximum value of magnetic flux density
$B_\parallel$	longitudinal component of magnetic flux density
$B_\perp$	transverse component of magnetic flux density
B_θ	azimuthal component of magnetic flux density
b	semi-minor axis of an ellipse
$\mathcal{C}^n$	class of differentiable functions
c	speed of light in a vacuum
D	maximal drift distance, see fig. 1.20
d	distance or width
$\boldsymbol{E}$	vector of electric field
$\boldsymbol{\mathcal{E}}$	a vector parallel to the electric field

E	electric field
E_{kin}	(electron) kinetic energy
$E_{\parallel}$	longitudinal component of electric field
$E_{\perp}$	transverse component of electric field
E_{θ}	azimuthal component of electric field
e	(positive) elementary charge
$\boldsymbol{F}$	force
$F(E_{\mathrm{kin}})$	cumulative distribution function of $f(E_{\mathrm{kin}})$
f_{α}	a special function
f_{c}	electron cyclotron frequency
$f(E_{\mathrm{kin}})$	probability density function of the spent beam kinetic energy spectrum
$G(\phi)$	transformation matrix for a two-dimensional counter-clockwise affine rotation over the angle ϕ
$G_n(u)$	value function of the dynamic programming algorithm
$\boldsymbol{H}$	vector of magnetic field strength
$\mathcal{H}$	Hamiltonian
$\boldsymbol{h}(z)$	function of a three-dimensional helical curve
h	height
I	current
I_{b}, $I_{\mathrm{b}n}$	current of electron beam (collected at the n'th stage)
I_{p}	current of primary electrons

I_{ref} or I_{reflect}	reflected (back-streamed) current of electron beam
I_{s}	current of secondary electrons
$\boldsymbol{J}$	vector of current density
J	current density
K, K_n	(the n'th) coefficient
$\mathcal{K}_\iota^n$	a special subset of $\mathcal{C}^n$
$\mathcal{L}$	Lagrangian
L	length
l	a variable for length, $l \in [0, L]$
m	mass
m_0	rest mass of an object
m_{e}	rest mass of an electron
N	number of voltage-depression stages
n	an integer
n_{p}	ion density in plasma
$P(u_1, u_2)$	power collected ideally by the stage with depression voltage u_1 (the next stage is has $U_d = u_2$)
P_{acc}	power of the accelerated electron beam
P_{beam}	power of spent electron beam
P_{output}	output power
P_{rec}	recovered (recuperated) power of spent electron beam

$p_{r,z}$	components of canonical momentum
p_θ	canonical angular momentum
p	(canonical) momentum
Q	fusion gain factor
q	charge
$\mathbb{R}$	the set of real numbers
$\boldsymbol{R}$	2D vector of electron (global) radius at a beam cross section
$\boldsymbol{R}_0$	2D vector of initial (global) radius of an electron at a beam cross section in the non-adiabatic transition model
R_0	length of $\boldsymbol{R}_0$
R_{max}	maximal reachable (global) radius of an electron
$\boldsymbol{r}$	position vector
$\boldsymbol{r}_0$	2D vector of initial Larmor radius at a beam cross section in the non-adiabatic transition model
r	radial coordinate or radius
r_0	initial radius
r_{b}	radius of a (linear) electron beam
r_{gc}	guiding center radius
$\boldsymbol{r}_{\text{L}}$	2D vector of the Larmor radius at a beam cross section
r_{L}	Larmor radius
s	(canonical) coordinate of an integral path
$\text{sgn}(x)$	-1 if $x < 0$ otherwise 1

T	plasma temperature
t	time
t_d	drift time
U	electric potential
U_b	beam voltage
U_body	gyrotron body voltage (potential)
U_col	gyrotron collector voltage (potential)
U_d, $U_{\mathrm{d}n}$	depression voltage (of the n'th collector stage)
u	the ratio between the electron total velocity V and its cyclotron frequency ω_c in the non-adiabatic transition model
$\boldsymbol{V}_\perp$	2D vector of initial transverse velocity at a beam cross section in the non-adiabatic transition model
v_θ	azimuthal velocity
$\boldsymbol{v}$	velocity
$\boldsymbol{v}_\mathrm{c}$	linear velocity of a circular (electron cyclotron) motion
$\boldsymbol{v}_\mathrm{d}$	drift velocity
V, $V_{z,r,\theta}$	initial electron velocity and its components in the non-adiabatic transition modeling
v	speed
v_0	initial longitudinal velocity
$v_\parallel$	longitudinal component of velocity
$v_\perp$	transverse component of velocity
x, y	transverse Cartesian coordinates

z	longitudinal (axial) coordinate
z_0	a special z position
Δ	discriminant
ΔU_d	difference between two depression voltages
ϑ	angle between electric and magnetic fields
Φ	effective potential
α	a coordinate in the Clebsch form of magnetic field
α_p	pitch factor
α_t	angle of the non-adiabatic transition model
β	a coordinate in the Clebsch form of magnetic field
γ	Lorentz factor
δ_1, δ_2	two special angular ranges, see fig. 3.26
δ_s	secondary electron yield
ε	permittivity
ζ	a variable related to z
η_col	collector efficiency
η_int	gyrotron interaction efficiency
η_RF	gyrotron RF efficiency
η_total	gyrotron overall (total) efficiency
θ	azimuth angle
θ_B	azimuth angle relative to $\boldsymbol{B}$ field

ι	a real positive parameter
λ	longitudinal period length of the electron gyrating trajectory
μ	permeability
μ_0	permeability of vacuum
μ_{mm}	electron magnetic moment
μ_{r}	relative permeability
$\boldsymbol{\nu}$	the normal vector of the collector wall
ρ	steepness
ρ_{q}	charge density
$\rho(r)$	a function depends only on r
σ	the ratio between the initial azimuthal speed V_θ and the total speed V in the non-adiabatic transition
τ	confinement time
ϕ	a variable for angle
φ	electric potential
$\varphi(z)$	a special function of z
ψ	magnetic flux
ψ_{L}	magnetic flux enclosed in a period of Larmor motion
$\psi(z)$	a special function of z
ω_B	frequency of magnetic field variation, seen by a particle
ω_{c}	electron cyclotron angular frequency

1 Introduction

1.1 Motivation

Gyrotrons are high-power microwave sources delivering Radio Frequency (RF) powers at megawatt (MW) level and at Continuous-Waves (CW). The possible operating frequency starts at around 30 GHz and ranges up to the terahertz (THz) frequency range.

High-power fusion gyrotrons are the only RF sources capable to deliver the required RF output power for the Electron Cyclotron Resonance Heating (ECRH) and Current Drive (ECCD) [1] in today's fusion experiments, e.g. Wendelstein 7–X (W7–X) stellarator [2], ITER tokamak [3] as well as in a future DEMOnstration fusion power plant. It is expected that DEMO will require a minimum injecting power of 50 MW for its ECRH and ECCD, hence it will require the simultaneous operation of numerous gyrotrons, each of which has to efficiently deliver an RF output power of higher than 1 MW at frequencies around or even above 200 GHz.

The fusion energy gain factor (often expressed as Q-factor [4]) depends directly on the efficiency of the plasma heating system. For the ECRH and ECCD systems, the efficiency is determined by the gyrotron efficiency. ITER is asking for a gyrotron overall (total) efficiency of $\eta_{\text{total}} \approx 50\,\%$. That will not be sufficient for the DEMO EC system of a 50 MW RF output power. Even an improvement of η_{total} from 50 % to 63 % would save 20 MW input electricity. Therefore, to meet the requirement of the future DEMO power plant, gyrotrons should have an overall efficiency which will be significantly higher than 60 %.

In order to improve the overall efficiency of a gyrotron, in the first place, the energy transfer (interaction) between the electron beam and the RF has to be optimized. But the increase of the interaction efficiency is very limited (e.g. an absolute increase of 3.2 % is shown in [5]). The gyrotron overall efficiency can be tens of percent points higher, if the energy recuperation mechanism at the collector (the collector has a so-called collector efficiency η_{col}) is able to regain more energy in the spent electron beam than with today's technology.

Today, only Single-stage Depressed Collectors (SDC) are used in gyrotron operations. However, applying an SDC and considering a typical interaction efficiency between 30 % and 35 % will limit the gyrotron overall efficiency below 60 %. The strong magnetic field in gyrotron collector causes the application of Multistage Depressed Collector (MDC) concepts to be challenging. All kind of existing research towards an effective MDC concept [6–12] could not be successfully validated until today. Hence, the investigation of advanced MDC concepts is the target of this work.

To find an effective MDC concept, a systematic study of all kinds of theoretical possibilities is performed. The two fundamental types of MDC concepts including their variants are systematically investigated. They are assessed according to a certain set of criteria. The most promising MDC design approach is proposed during this work [13–15]. This design can achieve the expected efficiency, while further optimizations allow it to be even several times compacter than the others. The mechanical design and assembling steps are given at the end of this monograph. With this MDC proposal, a gyrotron overall efficiency of above 60 % will be hoped for.

1.2 Gyrotrons for Nuclear Fusion

1.2.1 Principle of Nuclear Fusion

Nuclear fusion is a reaction in which two or more atomic nuclei form one or more different nuclei and subatomic particles. For example, the fusion of deuterium (D) and tritium (T) produces one neutron (n) and one non-radioactive helium:

$$ {}_{1}^{2}\mathrm{D} + {}_{1}^{3}\mathrm{T} \longrightarrow {}_{2}^{4}\mathrm{He}\ (3.5\,\mathrm{MeV}) + {}_{0}^{1}\mathrm{n}\ (14.1\,\mathrm{MeV}) \ . \tag{1.1} $$

The total rest mass of the matters decreases, while the reduction of the mass releases a large amount of energy. This new kind of energy source is very promising for the future, since fusion power plants will be capable of providing the grid base-load without polluting the environment nor causing global warming. A base-load power plant should be sufficiently efficient. This should also be shown in the first DEMOnstration fusion plant.

In order to activate fusion reactions, the positively charged nuclei should be brought so close, that the attractive nuclear force could be higher than the repulsive Coulomb force. However, it is not trivial to do that. One criterion is the temperature, which has to be at the order of 10 to 100 keV ($\sim 10^8$ to 10^9 K) [16]. At this energy, atoms cannot be formed, such that nuclei and unbounded electrons are in a plasma. An energy break-even for the D-T fusion has to fulfill the criterion [16, 17]

$$n_{\mathrm{p}}\, T\, \tau > 3 \times 10^{21}\,\mathrm{m}^{-3}\,\mathrm{keV\,s}\,, \tag{1.2}$$

where n_{p} and T are the ion density and temperature in the plasma, and τ is the energy confinement time. There are two approaches to achieve this break-even criterion. The first approach is called inertial confinement, where compressed pellets are heated by lasers and further compressed by the counterforce of explosions. The other approach is the magnetic confinement, where plasma species rotates around magnetic field lines. Tokamak and stellarator are the typical devices for the magnetically confined fusion.

The magnetically confined plasma should be heated to such a high temperature, that the nuclei in their random encounters would have enough kinetic energy to overcome the repulsion force. Heating the plasma to that temperature takes considerable energy. Highly efficient generation of this heating energy is a big challenge, which has to be solved for the DEMO power plant. As the ohmic heating is ineffective for the fusion plasma due to its low resistance at high plasma temperature, the heating power is injected either by the Neutral Beam (NB) or the RF heating system.

As one of the RF heating systems, the ECRH is successfully used as the main heating method in the W7–X stellarator. On the other hand, ECCD can drive the plasma current in a tokamak. In addition to the plasma heating and current drive, the EC system is able to deposit megawatts of power inside a localized volume. Therefore, it has been successfully applied in the stability control for the plasma breakdown, saw-teeth and neoclassic tearing modes [18]. The ECRH and ECCD are based on the electron cyclotron resonance in the magnetic field. Electrons rotate in a (locally homogeneous) magnetic field with the so-called electron cyclotron frequency

$$f_\mathrm{c} = \frac{\omega_\mathrm{c}}{2\pi} = \frac{e\,B}{2\pi\,\gamma\,m_\mathrm{e}} \approx \frac{B \cdot 28}{\gamma}\frac{\mathrm{GHz}}{\mathrm{T}} \tag{1.3}$$

where B is the magnetic flux density, e is the elementary charge, m_e is the rest mass of an electron and

$$\gamma = \left(1 - v^2/c^2\right)^{-\frac{1}{2}} \tag{1.4}$$

is the Lorentz factor [19]. In eq. (1.4), v is the speed of the electron and c is the speed of light in vacuum. Depending on the magnetic field and the γ of electrons (as well as a Doppler shift of the resonance frequency) at the target position, one can determine the required RF frequencies for ECRH and ECCD. For instance, the EC system of W7-X stellarator are operating at 140 GHz, the ITER tokamak requires a frequency of 170 GHz, while for the future DEMO fusion power plant, frequencies up to 240 GHz are foreseen [20].

The RF power is transmitted via either a Quasi-Optical (QO) system (e.g. W7-X) or transmission lines (e.g. ASDEX Upgrade [21]) to the plasma. Figure 1.1 shows the schematic of a general EC system for ITER using transmission lines (TL). The quasi-optical output RF beam from gyrotron is converted to a hybrid mode (HE_{11}) by the Matching Optics Unit (MOU) then fed to the TL. The TLs are evacuated corrugated waveguides such that the hybrid mode can propagate. Along TLs there are miter bends which turns the RF propagation direction, and polarizers which rotates the beam polarization. On the tokamak side of the TL there is a window (so-called torus window) separating the vacuum inside the TL and the tokamak. The RF beam is finally injected into the plasma through the EC launcher.

Technically, one can improve the efficiency of the total EC system by optimizing the efficiency of the RF generation and transmission. Increasing the EC efficiency by 10 % would save approximately 20 MW of input power for a 50 MW EC system like in DEMO, which would save millions of Euros each year for the cost of electricity, relax the cooling system and also reduce the total power supply costs [23].

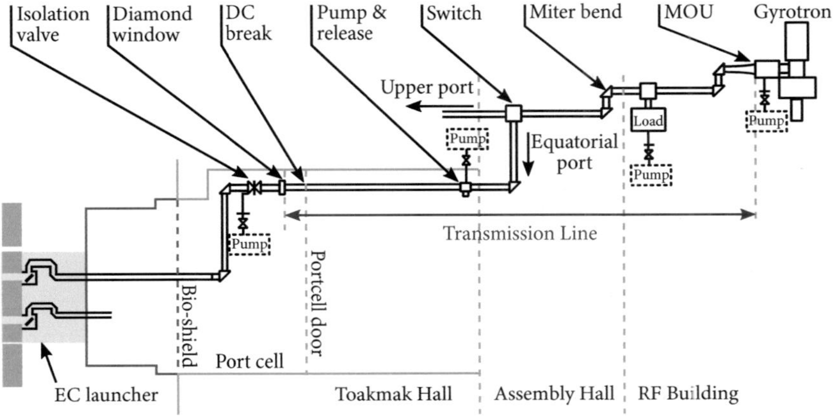

Fig. 1.1: Schematic of a general EC system for ITER [22]

The EC transmission system usually has an efficiency of 90 % to 92 % [22], whereas the gyrotrons using SDCs have approximately 50 % overall efficiency. Hence, a large margin in the efficiency of an EC system can be filled up by increasing the efficiency of the individual gyrotrons.

1.2.2 Fundamentals of Fusion Gyrotrons

The gyrotron converts the kinetic energy involved in the electron transverse gyration to the RF field. In gyrotron, electrons rotate in a magnetic field with a frequency given by eq. (1.3). If the magnetic flux density B is locally constant and homogeneous, the cyclotron frequency ω_c is constant in a non-relativistic scenario. However, if the Lorentz factor γ is taken into account, ω_c should depend on the Lorentz factor, which depends indirectly on the kinetic energy of a special electron. High-energetic electrons have high γ, thus low ω_c, while low-energetic electrons have low γ and high ω_c.

The gyration radius (so-called Larmor radius r_L) is

$$r_L = \frac{v_\perp}{\omega_c}, \tag{1.5}$$

where $v_\perp$ is the transverse velocity of the particle (electron). Another parameter related to the cyclotron motion is the pitch factor α_p, which is defined as

$$\alpha_\mathrm{p} = \frac{v_\perp}{v_\parallel}, \tag{1.6}$$

where $v_\parallel$ is the longitudinal velocity of the particle. For example, α_p is larger than 1.2 in the cavity (where the interaction takes place) of the 170 GHz 1 MW gyrotron [24] and approximately 1.3 in the 170 GHz 2 MW coaxial cavity gyrotron [25].

Any RF wave propagating in a (cylindrical) waveguide can be decomposed to a series of Transverse Electric (TE) and Transverse Magnetic (TM) modes. In a TE-mode, the electric field only has transverse components. This transverse RF electric field can accelerate or decelerate electrons transversely, while the influence of the RF magnetic field on the electrons is negligible in TE-modes. It follows an energy exchange (interaction) between RF field and the electron gyration motion. Considering an ideal initial electron beam, where all electrons have the same kinetic energy, while the same portion of the total energy are in the transverse motion of each electron; those electrons which gained energy from the RF field become faster and "heavier", therefore, they have larger r_L and smaller ω_c (lagging in the gyration phase); whereas electrons who transferred energy to the RF field have higher ω_c (coming up in the gyration phase) and smaller r_L. In this way, electrons are gradually bunched. If the RF field and the electron gyration have the same frequency, there will be no net energy exchange. If the angular frequency of RF field is slightly higher than the initial ω_c, the electrons are gradually decelerated by the electric field. As the velocity $v_\perp$ (and γ) is decreasing, the gyration frequency becomes higher. In this case there is a net energy transfer from the electrons to the RF field. This interaction takes place in the gyrotron cavity. The cavity has a series of resonant modes, each of which has a different eigen frequency. In the operation, the desired mode is selected by accurately placing the electron beam on the appropriate global radius.

Figure 1.2 shows the schematic of a gyrotron with a conventional hollow cavity. The gyrotron main Super-Conducting (SC) magnet creates a strong magnetic field (e.g. higher than 6 T in the 170 GHz 1 MW gyrotrons) to achieve the desired EC frequency ω_c.

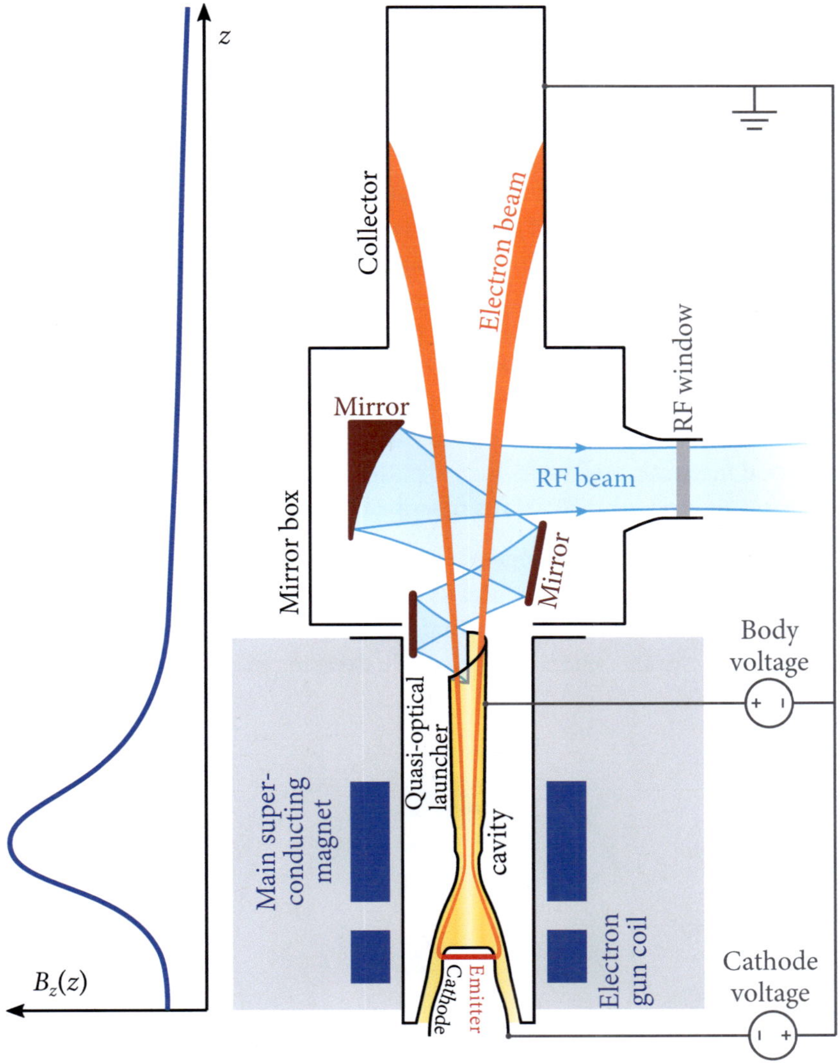

Fig. 1.2: Schematic of a gyrotron

Electrons are emitted by the emitter ring. They form an annular electron beam. The cross section of the annular electron beam is shown in fig. 1.3. In fusion gyrotrons, the Larmor radius r_L is much smaller than the radius of any electron guiding center r_{gc}. Therefore, this electron beam is called a small-orbit beam. The emitted electrons are accelerated by a voltage in tens of kV (e.g. up to 80 kV in the 170 GHz 1 MW gyrotron [26]), to achieve the target γ and transverse velocity $v_\perp$. More than half of their kinetic energy will be in the transverse gyration when they reach the cavity. The electron trajectories follow magnetic field lines (to be accurate, they follow the magnetic flux surfaces). On the path to the gyrotron cavity, the orbits and guiding center radii of the electrons are compressed as the magnetic field becomes higher, due to the conservation of magnetic moment and canonical angular momentum, respectively (see later in section 1.3.1). The cavity is located at the maximum of the magnetic field, where the local magnetic field varies only slightly. In the cavity, the electron beam transfers energy to the RF field as described in the previous paragraph.

Next to the cavity, the quasi-optical launcher converts the TE-mode to a quasi-optical mode, then the quasi-optical RF beam is output horizontally through the RF window, as shown in fig. 1.2. On the other hand, the spent electron beam leaves the cavity and enters the region of a decaying magnetic field.

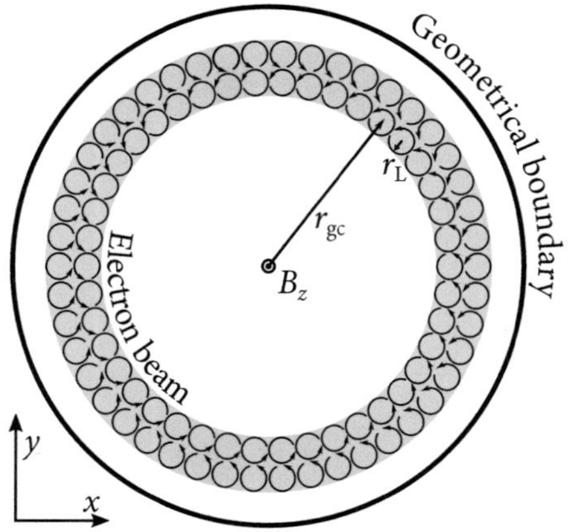

Fig. 1.3: Cross section of the annular electron beam in a small-orbit gyrotron

Even in the decaying magnetic field, the electron beam still follows the magnetic field lines. The beam cross section keeps the small-orbit shape as qualitatively shown in fig. 1.3. However, the electron guiding center radii and the beam thickness are expanded in the decaying field. The spent beam electrons are collected at the collector eventually. The remained kinetic energy of electrons is converted to heat and unwanted electromagnetic waves (including X-rays), or carried out by the emitted secondary electrons.

1.2.3 Theoretical Estimation of the Gyrotron Efficiency

The overall efficiency of a gyrotron is the ratio between the output power of the quasi-optical RF beam and the total DC input power. It is determined by the efficiencies of three sub-systems.

In the interaction, most electrons donate their kinetic energy of the cyclotron motion partly to the RF field. There is still a significant energy remained in the cyclotron motion. Moreover, the energy in the electron longitudinal motion is untouched. These correspond to a remained beam power of higher than 2 MW in the case of the 170 GHz 1 MW gyrotron. Besides, there can be a small amount of electrons, which gain energy from the RF field, because they did not enter the field in proper phases. The interaction efficiency (also called electron efficiency) η_{int} is the ratio between the generated RF power and the power of the accelerated electron beam. Fusion gyrotrons have a typical interaction efficiency between 30 % and 35 % [27].

The RF efficiency η_{RF} is the ratio between the output RF power and the generated one. The assumption of $\eta_{\mathrm{RF}} \approx 90\,\%$ (can be found e.g. in [5, 28]) includes different kinds of RF losses of the quasi-optical launcher, mirrors and the window.

When the spent electron beam is collected at the collector, the energy carried by the electrons is considered as loss in the electric circuit. With the technique explained in section 1.4.1, a part of the spent beam energy can be recuperated. The collector efficiency

$$\eta_{\mathrm{col}} = \frac{P_{\mathrm{rec}}}{P_{\mathrm{beam}}} \tag{1.7}$$

is the ratio between the recuperated (recovered) electronic energy (P_{rec}) and the energy of spent electron beam (P_{beam}). It is zero, if no recuperation is applied. Nowadays gyrotrons can have η_{col} up to 60 % using SDCs.

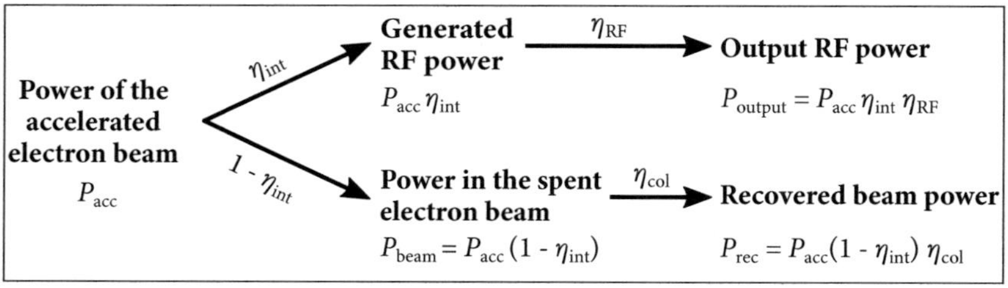

Fig. 1.4: Relation between power and efficiencies in a gyrotron

The relation between powers and efficiencies in a gyrotron is presented in fig. 1.4. The expression of the overall gyrotron efficiency

$$\eta_{total} = \frac{P_{output}}{P_{acc} - P_{rec}} = \frac{\eta_{int}\,\eta_{RF}}{1 - (1 - \eta_{int})\,\eta_{col}} \tag{1.8}$$

is derived from that diagram. Equation (1.8) is visualized in fig. 1.5a. Increasing any of the three efficiencies will raise the gyrotron total efficiency. The interaction efficiency could be further increased up to 43 % when the gyrotron works in a “hard-self-excitation” region [5]. However, there is only little margin for the increment of the interaction efficiency, whereas the collector efficiency can be improved by tens of percent points with advanced multistage collectors. Furthermore, the first and second derivative of η_{total} to η_{col}

$$\frac{\partial \eta_{total}}{\partial \eta_{col}} = \eta_{RF} \frac{(1 - \eta_{int})\,\eta_{int}}{\left[1 - (1 - \eta_{int})\,\eta_{col}\right]^2} > 0\,, \tag{1.9a}$$

$$\frac{\partial^2 \eta_{total}}{\partial {\eta_{col}}^2} = \eta_{RF} \frac{2\,(1 - \eta_{int})^2\,\eta_{int}}{\left[1 - (1 - \eta_{int})\,\eta_{col}\right]^3} > 0\,, \tag{1.9b}$$

are both positive. The positive derivatives mean that as η_{col} increases, it will have a higher contribution to the gyrotron total efficiency. This is shown in fig. 1.5b. With a relatively low η_{col}, e.g. on the white curve labeled with 1 : 2, increasing 2 % of η_{col} (absolutely) will cause η_{col} to be increased by only 1 %

(absolutely); whereas with a higher η_{col}, like on the red "1 : 1" curve, 1 % increment of η_{col} corresponds to 1 % increment of η_{total}. In the extreme case, if there would be a collector with $\eta_{col} = 100\,\%$, it would totally compensate the lack of interaction efficiency.

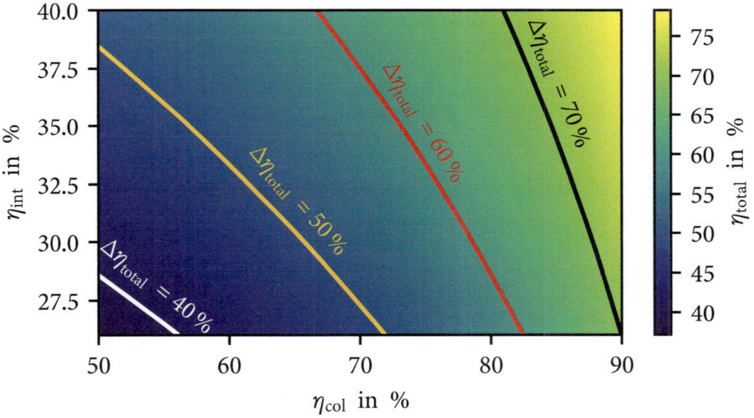

(a) Visualization of eq. (1.8)

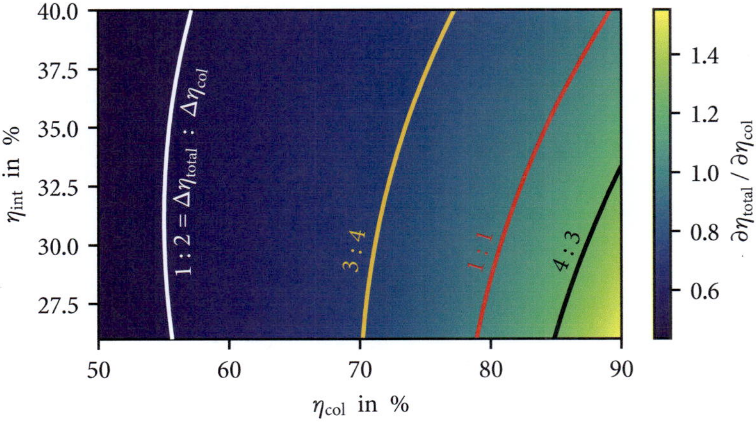

(b) Visualization of eq. (1.9a)

Fig. 1.5: Relation of η_{int}, η_{col} and η_{total} for $\eta_{RF} = 90\,\%$

Additionally, it can be read from fig. 1.5a that to achieve the goal of $\eta_{\text{total}} > 60\,\%$ (assuming $\eta_{\text{int}} = 35\,\%$), η_{col} has to be greater than 74 %. Such a high η_{col} exceeds the capability of the existing gyrotron collectors. Thus, the goal of this work is to deliver an advanced collector concept, which is capable to have an η_{col} higher than 74 %. None of the existing collector proposals for gyrotrons could even theoretically achieve this high efficiency.

1.3 Basic Physical Principles

1.3.1 Electron Motion in Electrostatic and Magnetostatic Fields

Lorentz Force and *E*×*B* Drift

The Lorentz force

$$\boldsymbol{F} = q(\boldsymbol{E} + \boldsymbol{v} \times \boldsymbol{B}) \tag{1.10}$$

arises from the space-time averaged forces acting on material charges and current [29]. Equations of the electron motion in gyrotrons can be derived from the Lorentz force. There are three important phenomena from the motion equations: $\boldsymbol{E}\times\boldsymbol{B}$ drift, adiabatic invariant and Busch's theorem. They are the basics for the gyrotron collector design and thus will be introduced in this section.

Equation (1.10) can be decomposed to the projection on the direction of $\boldsymbol{B}$ and the one perpendicular to the $\boldsymbol{B}$-field. The electric field component $\boldsymbol{E}_\parallel$ parallel to $\boldsymbol{B}$ causes a force of straight acceleration or deceleration

$$\frac{1}{q}\boldsymbol{F}_\parallel = \boldsymbol{E}_\parallel\,, \tag{1.11}$$

whereas the force on the plane perpendicular to $\boldsymbol{B}$ is

$$\frac{1}{q}\boldsymbol{F}_\perp = \boldsymbol{E}_\perp + \boldsymbol{v}_\perp \times \boldsymbol{B}\,. \tag{1.12}$$

Let $\boldsymbol{v}_\perp$ be the sum of two velocities $\boldsymbol{v}_{\text{c}}$ and $\boldsymbol{v}_{\text{d}}$:

$$\boldsymbol{v}_\perp = \boldsymbol{v}_{\text{c}} + \boldsymbol{v}_{\text{d}}\,, \tag{1.13}$$

then eq. (1.12) can be written as

$$\frac{1}{q}\boldsymbol{F}_{\perp} = \boldsymbol{v}_{\mathrm{c}} \times \boldsymbol{B} + (\boldsymbol{E}_{\perp} + \boldsymbol{v}_{\mathrm{d}} \times \boldsymbol{B}) \; . \tag{1.14}$$

It will be shown in this section that $\boldsymbol{v}_{\mathrm{c}}$ is the linear velocity of the electron cyclotron motion, while the Ansatz of the drift velocity $\boldsymbol{v}_{\mathrm{d}}$ is

$$\boldsymbol{v}_{\mathrm{d}} = \frac{\boldsymbol{E} \times \boldsymbol{B}}{B^2} \; . \tag{1.15}$$

Substituting eq. (1.15) into the parenthesis of eq. (1.14), one gets

$$\boldsymbol{E}_{\perp} + \boldsymbol{v}_{\mathrm{d}} \times \boldsymbol{B} \; = \; \boldsymbol{E}_{\perp} + \frac{(\boldsymbol{E} \times \boldsymbol{B}) \times \boldsymbol{B}}{B^2} \; = \; \boldsymbol{E}_{\perp} + \frac{(\boldsymbol{E}_{\perp} \times \boldsymbol{B}) \times \boldsymbol{B}}{B^2} \; = \; 0 \; . \tag{1.16}$$

Equation (1.16) vanishes since these two cross products actually mean that the $\boldsymbol{E}_{\perp}$ vector is rotated around $\boldsymbol{B}$ by 90°, twice, which cancels the former $\boldsymbol{E}_{\perp}$ term.

As $(\boldsymbol{E}_{\perp} + \boldsymbol{v}_{\mathrm{d}} \times \boldsymbol{B})$ vanishes, eq. (1.14) becomes

$$\frac{1}{q}\boldsymbol{F}_{\perp} = \boldsymbol{v}_{\mathrm{c}} \times \boldsymbol{B} \; . \tag{1.17}$$

This is the periodic cyclotron motion perpendicular to the magnetic field. It has the Larmor radius

$$r_{\mathrm{L}} = \frac{m\, v_{\mathrm{c}}}{q\, B} \; . \tag{1.18}$$

Equation (1.15) is the so-called **E**×**B** drift. The drift velocity $\boldsymbol{v}_{\mathrm{d}}$ is independent of the mass and (even the direction of) the velocity $\boldsymbol{v}$ of a particle. This is an important phenomenon for the design of advanced collectors and will be applied in chapter 3.

Adiabatic Invariant

According to classical mechanics, if there is a periodic motion, the action

$$\oint p \, \mathrm{d}s$$

is constant under certain circumstances, where p is a momentum and ds is a piece of motion path in the projected plane of the periodic motion. This constant action is called adiabatic invariant.

There are three adiabatic invariants for free electron motions. The most important one for the collector design is the conservation of the magnetic moment. The integral of the angular momentum is

$$\oint p\,\mathrm{d}s = \oint m\,v_\mathrm{c}\,r_\mathrm{L}\,\mathrm{d}\theta_B = 2\pi\,r_\mathrm{L}\,m\,v_\mathrm{c} = 4\pi\frac{m}{|q|}\mu_\mathrm{mm}\,, \tag{1.19}$$

where

$$\mu_\mathrm{mm} = \frac{m\,v_\mathrm{c}^2}{2\,B} \tag{1.20}$$

is the magnetic moment. The magnetic moment μ_mm should be constant, if the magnetic field varies inconspicuous (both in time and space) within an electron cyclotron period, and the ratio m/q does not change. Inserting eq. (1.18) into eq. (1.20) there is

$$\mu_\mathrm{mm} = \frac{q^2}{2m}\,B\,r_\mathrm{L}^2 = \frac{1}{2\pi}\,\frac{q^2}{m}\,\psi_\mathrm{L} \tag{1.21}$$

as far as $\omega_B \le \omega_\mathrm{c}$ [30], where ψ_L is the magnetic flux (non-strictly) enclosed in the electron orbit and ω_B is the frequency of the magnetic field as seen by the (moving) particle.

The magnetic field has a maximum at the center of the gyrotron cavity. After the center of the cavity, the electrons travel into a decaying magnetic field. According to eq. (1.20), reducing B means that the cyclotron velocity v_c [as the major part of $v_\perp$, see eq. (1.13)] is decreasing. The kinetic energy in the original transverse motion is converted to the longitudinal motion in the adiabatic decaying magnetic field. In a conventional collector, one can calculate v_c and r_L of an electron at different axial positions according to the adiabatic invariant.

The adiabatic decaying of magnetic field has a beneficial effect in the design of a collector. As the energy of the transverse motion is converted to the longitudinal motion, this part of energy becomes also possible to be recuperated. However, the adiabatically decaying of magnetic field requires a long axial distance and it leads to a large collector.

Busch's Theorem

In an axisymmetric system, the Lagrangian of a charged particle is [31]

$$\mathcal{L} = -m_0\, c^2 \sqrt{1 - \frac{|\boldsymbol{v}|^2}{c^2}} - q\,(U - \boldsymbol{A} \cdot \boldsymbol{v})\,, \tag{1.22}$$

where $\boldsymbol{v} = (\dot{r}, r\dot{\theta}, \dot{z})$ is the particle velocity, $\boldsymbol{A}$ is the magnetic vector potential and U is the electric potential. If $\boldsymbol{A}$ and U are symmetric around the axis (where $\boldsymbol{A}$ only has the azimuthal component) i.e. $\partial \boldsymbol{A}/\partial\theta$ and $\partial U/\partial\theta$ are zero, the azimuthal (θ) part of the Lagrangian equation [32]

$$\frac{\mathrm{d}}{\mathrm{d}t}\left(\frac{\partial\mathcal{L}}{\partial\dot{\theta}}\right) - \frac{\partial\mathcal{L}}{\partial\theta} = 0 \tag{1.23}$$

remains

$$\frac{\partial\mathcal{L}}{\partial\dot{\theta}} = p_\theta = \gamma m_0\, r^2\, \dot{\theta} + q\, r\, A_\theta = \gamma m_0\, r^2\, \dot{\theta} + \frac{q}{2\pi}\psi = \text{const.}\,, \tag{1.24}$$

where ψ is the enclosed flux in the axisymmetric cross section with the radius r. Equation (1.24) indicates the conservation of the canonical angular momentum p_θ. This is Busch's theorem [33].

A special case of eq. (1.24) is the conservation of the magnetic flux at the specific moments of each cyclotron period when the angular velocity $\dot{\theta}$ vanishes, that is

$$\psi = \text{const.} \quad \text{when } \dot{\theta} = 0\,. \tag{1.25}$$

Figure 1.6 visualizes this equation. The cyclotron motions of two electrons, which have the same guiding center radius r_{gc}, are shown in this figure. The cyclotron motions enclose different magnetic fluxes $\psi_{\mathrm{L}①}$ and $\psi_{\mathrm{L}②}$. Their values will not change, if the adiabatic condition is fulfilled. On the other hand, the electrons have zero azimuthal velocity at those phases marked by the red dots. The edge of the yellow circle passes through the zero-v_θ points of the electron ②. The magnetic flux enclosed in that yellow circle is always constant, due to Busch's theorem in eq. (1.24). Since r_{gc} is typically one hundred times of r_{L}, the Larmor radius compared to r_{gc} is so small such that it can be approximately treated as:

$$r_{\mathrm{gc}} \approx r_{①} \approx r_{②}\,.$$

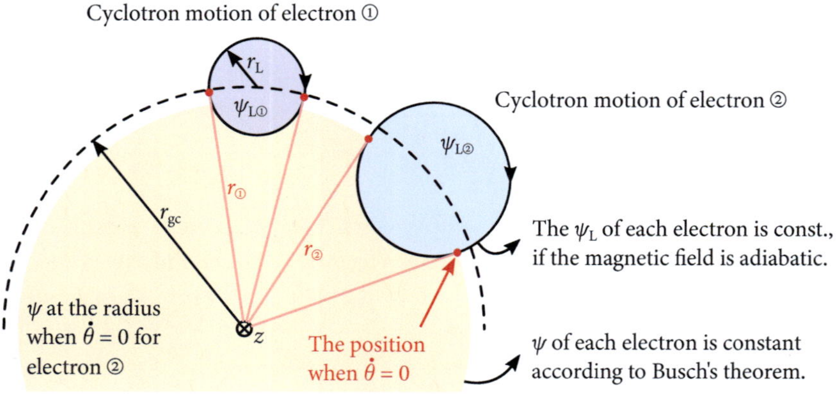

Fig. 1.6: Schematic of the electron beam cross section showing the magnetic fluxes in eqs. (1.21) and (1.25)

No matter how the electric and magnetic fields are, as far as the axisymmetric conditions for Busch's theorem ($\partial \boldsymbol{A}/\partial\theta = 0$ and $\partial U/\partial\theta = 0$) are fulfilled, all electron orbits will have very close guiding center radii, such that it is challenging to rearrange or separate electron orbits radially.

Even though both eqs. (1.21) and (1.25) show the properties about the conservation of magnetic fluxes, their prerequisites are different. The flux ψ_L of each electron orbit (see fig. 1.6) is constant only when the magnetic field varies adiabatically, whereas the conservation of flux ψ in the gyrotron hollow electron beam requires only the symmetry. The latter is even valid for a non-adiabatic magnetic field e.g. with a large ∇B_z, or $B = 0$ locally in a cusp electron gun [34].

1.3.2 Emission and Suppression of Secondary Electrons

When an electron with sufficient energy hits the surface of a solid, the surface emits electrons. The emitted electrons are called secondary electrons, while that original one is called primary electron. There are different kinds of secondary electrons and multiple sets of terminologies for categorization of secondary electrons in the literature. The most commonly referred categorization [35] classifies secondary electrons in three types:

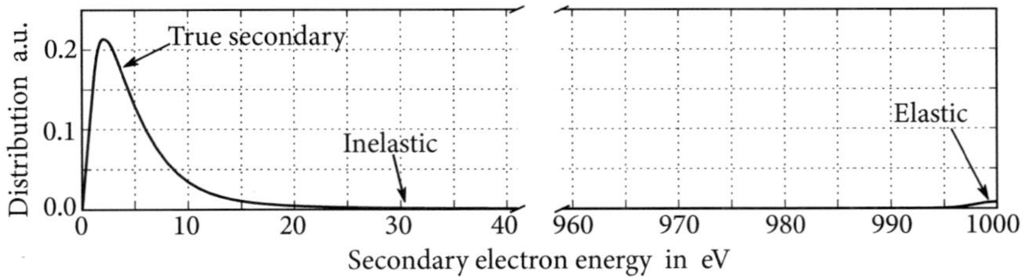

Fig. 1.7: Model of secondary electron energy distribution by a perpendicular hit of a 1 keV primary electron

- True secondary electrons are the slow ones with energies in the order of 10 eV. Their energies are nearly independent of the primary electron energy.
- An elastically reflected electron has only one collision with the atom. It is scattered elastically without energy loss.
- Inelastically reflected electrons have energies between the true secondaries and the elastically reflected ones. They arise from one or more inelastic collisions.

The investigation of secondary electrons in a gyrotron collector will be based on this categorization. The term *secondary electrons* in the following text also includes the elastically reflected ones. An example of the secondary electron energy distribution is shown fig. 1.7. This distribution is calculated from the secondary emission model [36, 37] for the case, when a primary electron with 1 keV energy impacts on a copper surface. This model is implemented in the simulation software ESRAY [38] and CST STUDIO [39, 40].

The ratio between the emitted current of total secondary electrons (I_s) and the impacting current of primary electrons (I_p):

$$\delta_s = \frac{I_s}{I_p} \tag{1.26}$$

is called Secondary Electron Yield (SEY). The SEY increases with the flatness of the primary electron impacting angle, while it is also related to the energy of the primary electron.

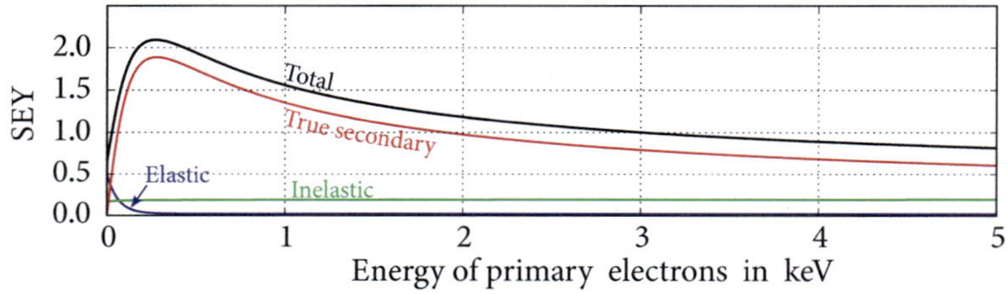

Fig. 1.8: SEY curve for perpendicularly impacted electrons

Figure 1.8 shows an example of the SEY and its compositions for different primary electron energies. In fig. 1.8 the maximum SEY can be higher than a factor of two for the primary electron energy of approximately 300 eV. Both SEY of the true secondaries and the total SEY decrease notably when the energy of primary electron increases. Starting from 3 keV, the total SEY in this example drops below unity, which means that the secondary electron current will be lower than the primary current. However, fig. 1.8 is only for a single generation of secondary electrons. In gyrotron collectors, trajectories of emitted secondary electrons may be affected by the magnetic field and the local electric field, such that some secondary electrons turn back to the surface where they were emitted. The surface under the bombardment again emits another generation of secondary electrons. Then, the second generation causes the third generation and so on. Finally, if one counts the total emitted current of all generations of the secondary electrons, there is no doubt that the gross emitted current is higher than the single-impacting SEY in fig. 1.8. Therefore, in the numerical simulation of secondary electrons, multiple generations of emission have to be taken into account. Besides, the sampling of the curve in fig. 1.7 will need numerical emissions of multiple macro electrons (see section 1.6) in random energy and angles according to the energetic and angular distribution, respectively. Hence, a very large number of secondary electron samples has to be taken into account in the numerical simulations, as will be discussed later.

Although, there has been a lot of researches since the beginning of the 20th century e.g. [41, 42], where the properties of secondary electron have been measured, the check of secondary electron influences in vacuum tubes was exclusively based on numerical simulations. In principle, numerous secondary electrons should be sampled (emitted) for each bombarding electron in each generation, and an infinite number of generations should be considered, until the simulation reaches a steady state. However, this process would lead to an exponential growth on the number of total macro electrons in the simulation, which is still impossible for today's high-performance computers. Therefore, only limited number of secondary electrons per bombardment is considered in realistic simulations. In addition, the maximum generation is limited for a reasonable approximation. As a compromise between accuracy and computational resource, the work in the 1990s e.g. [10] considered only a selected fraction of the secondary electrons due to the computer performance at that time. At the time of writing, the default secondary electron model in CST Studio samples maximally ten secondary electrons per bombardment and it has by default no limitation on the maximum generation [39], whereas the optimal secondary electron emission settings for the two-dimensional axisymmetric simulations in EsRay [38] are still under investigation. Since secondary electrons are generated in the simulation randomly according to their angular and energetic distribution, the randomness introduces a certain noise to the system. The steady state is considered as achieved, when the noise between simulation iterations is typically below −30 dB (see the figure of convergence in [14]).

Secondary electrons are harmful for the gyrotron operation for two reasons:

1. They could form a current between different potentials. Such an inter-potential current usually flows in the opposite direction to the primary current and causes power loss. This loss decreases the gyrotron efficiency.

2. Secondary electrons may be accelerated backwards from the collector. They will probably participate in the interaction again and could cause a reduction of the RF power.

In both cases, the secondary electrons may impact some places, where there was no electron bombardment expected. This could also be critical for the design.

The effective SEY can be reduced in two ways. The first option is related to the secondary emission coefficient of the material. The second one is to create textures on the material surface. Both technologies can be combined to further reduce the effective SEY.

From the aspect of material properties, copper is the common material used as electrode for its electrical and thermal conductivity. But copper has a higher secondary emission coefficient than a lot of materials, such as beryllium, pyrolytic graphite, soot, titanium carbide, and tantalum [43]. Moreover, copper used for the construction can have oxide and contaminant, they will cause a higher coefficient than with the ideal pure copper. Well baked-out copper has a lower SEY than a non-baked one [44]. The SEY of electrode surfaces can be reduced by a superposition layer of small particles [35]. Besides, a high dose of impacting electrons can also reduce the SEY, as far as the surface is kept inside a vacuum environment [44].

Both micro and macroscopic textures can trap secondary electrons. Oxidizing the copper at elevated temperature in air produces often cracks on the surface, such that the surface roughness increases [44]. Alternatively, ion sputtering can bring rough textures on the surface, too [43]. A rough surface has a lower SEY than a smooth one. Grooving the surface macroscopically also reduces the effective SEY significantly [35, 45, 46]. The latter will be applied in sections 2.5.2 and 3.5 as the only technique to reduce the effective SEY of copper surfaces.

In spite of reducing the SEY, there are other ways to suppress secondary electrons. In other particle beam collection devices like Faraday Cups (FC), emissions of secondary electrons are also unwanted. However, solutions which involve transverse electric fields in FCs [47] are not directly adoptable in gyrotrons because of existence of a strong magnetic field in the gyrotron collector. On the other hand, the magnetic field can be used to create an $\boldsymbol{E}\times\boldsymbol{B}$ drift, which is useful for the re-collection of the secondary electrons, this will be presented in chapter 3.

In this work, the simulations with secondary electrons are based on the assumption of the emission model for copper [36, 37]. The exact parameters of secondary emissions depend also on many other properties like the composition of the material, purity of the surface, smoothness, etc. Hence, the applied model should be calibrated by experiments. No experiment has been done to calibrate the secondary emission model, in particular, within a moderated magnetic field (of 10–100 mT). In this work, the assumed "standard copper" model can at least provide a qualitative estimation for the influence of secondary electrons in gyrotron collectors.

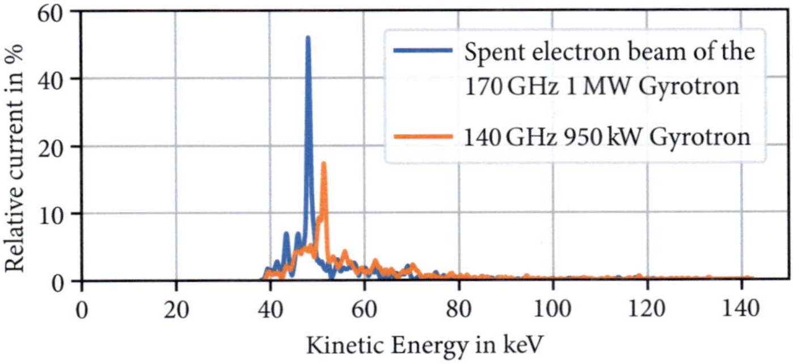

Fig. 1.9: Spent electron beam spectra of the 140 GHz gyrotron for W7-X and the 170 GHz 1 MW gyrotron for ITER

1.4 Technical Principles of Energy Recuperation in Vacuum Tubes

1.4.1 Basic Principle of Depressed Collectors

Spent Electron Beam

It has been explained in section 1.2.2, that during the interaction in the gyrotron cavity some electrons transfer energy to the RF field, while a minor part of beam electrons may gain kinetic energy from RF wave. The resulting spent electron beam after the interaction has a broad energy spectrum. For the W7–X and ITER gyrotrons developed at KIT, which have up to 80 kV accelerating voltage and 45 A beam current, their spent electrons could have kinetic energies in the range from 38 keV up to 130 keV. The integral (area) of the Probability Density Function (PDF) $f(E_{\mathrm{kin}})$ over the E_{kin} axis is unity. To assign the PDF a physical meaning, it can be interpreted as a relative current, such that the expectation of $f(E_{\mathrm{kin}})$ is proportional to the beam power. The energy spectra of two different gyrotron spent beams are shown in fig. 1.9. This realistic spent beam information is calculated from the self-consistent interaction code EURIDICE [48] and is one of the input parameters for the collector design. The beam power is

$$P_{\mathrm{beam}} = \frac{I_{\mathrm{b}}}{e} \int_0^{\infty} E_{\mathrm{kin}} f(E_{\mathrm{kin}})\,\mathrm{d}E_{\mathrm{kin}}\,, \tag{1.27}$$

where I_b is the total beam current and the integral is the expectation of the spectrum. If no action is taken, all power in eq. (1.27) will be finally absorbed at the collector, producing megawatts of power load at the collector wall. Depressed collectors can recuperate a part of the power and reduce the power load.

Single-Stage Depressed Collectors

Physically, a depressed collector decelerates the electrons before collecting them. In this way, the kinetic energy of each electron is partly converted back to the potential energy delivered by the high-voltage power supply. Technically, it means that the high-voltage power supply (the one provides the cathode voltage in fig. 1.2) delivers less power to the circuit without reducing the output RF power.

A Single-stage Depressed Collector (SDC) decelerates the entire electron beam by only one voltage. An ideal deceleration scheme for the spent beam spectrum of the 170 GHz 1 MW (ITER) gyrotron is presented in fig. 1.10. The power recovered from the deceleration of the electron beam is

$$P_{\text{rec}} = U_{\text{d}} I_{\text{b}} \tag{1.28}$$

where U_d is the depression (deceleration) voltage, namely the difference between the electric potential of collector body U_{body} and collector U_{col}.

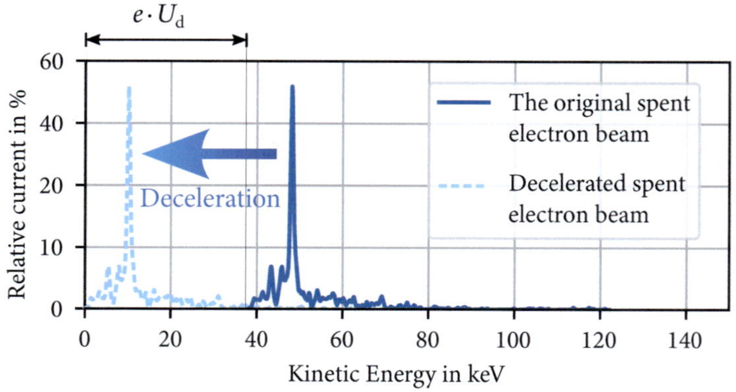

Fig. 1.10: Deceleration of the electron beam in an ideal SDC

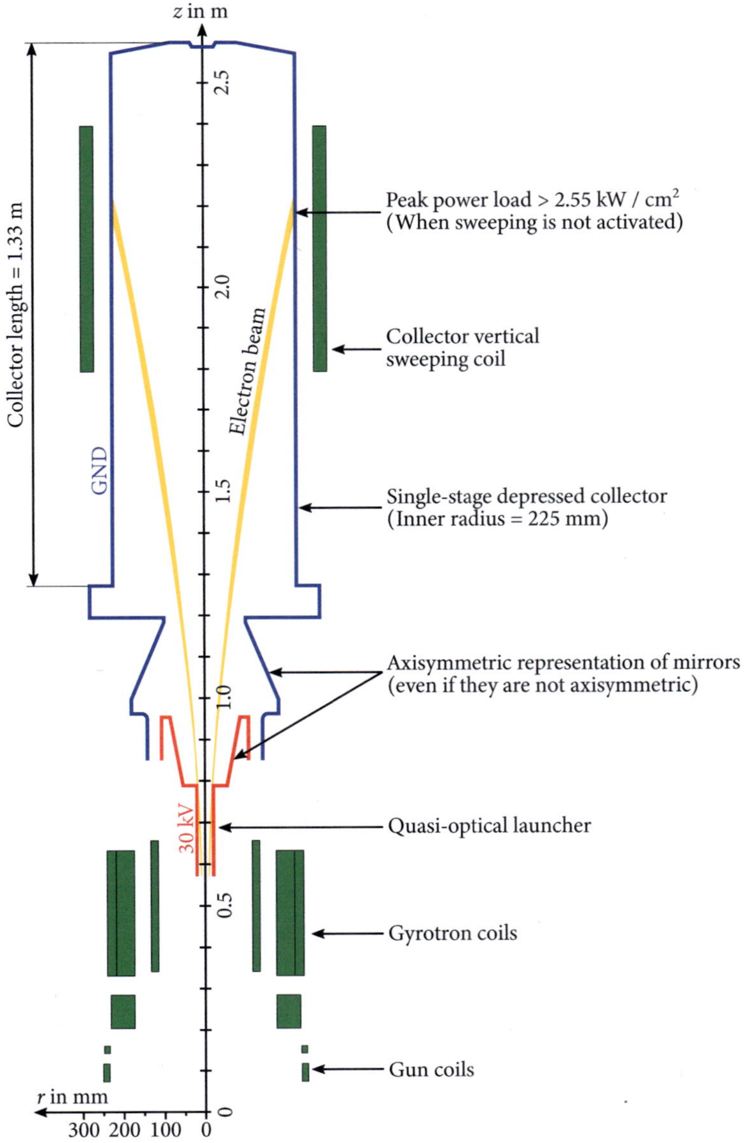

Fig. 1.11: The reference SDC for the 170 GHz 1 MW gyrotron

Assuming that electrons only have longitudinal velocities, such that they move exactly parallel against the electric field, the ideal depression voltage should just totally decelerate the slowest electron in the electron beam ($e \cdot U_\mathrm{d} = \min\{E_\mathrm{kin}\}$). However, the ideal depression voltage (approximately 38 kV in the example of fig. 1.10) cannot be achieved in practice. There are two important reasons for that:

1. According to eq. (1.20), the energy in the transverse motion is converted to the longitudinal motion as the magnetic field decays after the cavity. Although the maximal pitch factor of electrons is very low in the collector (typically, $\alpha_\mathrm{p} < 0.2$), there is still energy in electron transverse motion and $v_\perp \neq 0$. The depressed collector cannot recuperate this energy. If the ideal depression voltage, which corresponds to the total kinetic energy of an electron, was applied, the electron would be reflected due to the lack of longitudinal velocity to overcome the deceleration.
2. An electron induces positive charges on the geometrical boundaries, thus, there is also an attractive force between them. This separation of charges causes an additional potential energy to the electron in vacuum. If an electron comes closer to the metal wall, the attractive force yields a positive work, which means that this kind of potential energy is converted to the electron kinetic energy. When an electron beam leaves the gyrotron launcher into the mirror box, the distance from beam to the wall suddenly increases. Therefore, the beam electrons temporarily lose kinetic energy (up to 10 keV) and gain the same amount of the potential energy (if the radiations are neglected). The electrons will gradually gain the kinetic energy back, when they approach the collector electrodes.

In practice, less than 30 kV depression voltage [49] is applied to the spent beams in fig. 1.9. Recently in the experiment of a beam re-accelerating technique [49] (however, for another energy spectrum than fig. 1.9), the depression voltage can be even 38 kV. Nevertheless, to have a common base for the performance comparisons, $U_\mathrm{d} = 30$ kV (of the first stage, if there are multiple stages) is chosen in the simulations of the concepts presented in this work.

Since the coaxial-cavity DEMO gyrotron is still under development, the investigations are based on the existing 170 GHz 1 MW gyrotron. Whether the cavity is coaxial or conventional does not affect the principle of the collectors. Figure 1.11 shows the two-dimensional axisymmetric model of the SDC for the 170 GHz 1 MW gyrotron. It is the reference depressed collector in this work. The maximal radius of the inner wall and the length of the other collector designs for the 170 GHz 1 MW gyrotron will be based on the parameters of this SDC. Since the electron beam is confined by the magnetic field, the impacting area on the collector wall is small, such that the beam causes a local power load of higher than 2.55 kW/cm^2. The final effective power load will be averaged in time, when the impacting position of the electron beam on the collector is swept on a larger area via the sweeping coils. The temporally averaged power load should be less than 500 W/cm^2 [50]. In reality, the mirrors are not axisymmetric, but they are simplified in this two-dimensional modeling. The mirror box and the collector are grounded, while the launcher and the first mirror are at $U_{\text{body}} = 30\,\text{kV}$. A realistic spent electron beam calculated from EURIDICE is injected at the axial position of the launcher. As the beam approaches the collector, the magnetic field is weaker, and the annular beam becomes thicker.

As mentioned earlier, the depression voltage of an SDC cannot even fully decelerate the slowest electron in the spent electron beam, while the energy recuperation of the initially fast electrons is even less effective under such a depression voltage limited by the slow electrons. Hence, the collector efficiency (η_{col}) of an SDC is typically below 60 % (e.g. 57.4 % for the given energy spectrum of the 170 GHz 1 MW gyrotron), which renders a gyrotron overall efficiency (η_{total}) of approximately 50 %. To recover more power from the spent electron beam, an upgrade from single depression stage to multiple depressed stages is required.

Multistage Depressed Collectors

A Multistage Depressed Collector (MDC) splits the electron beam and sorts the electrons into multiple energy intervals according to their energy. Electrons which belong to different energy intervals will be decelerated by different depression voltages. In this way, the energy in the spent electron beam can be optimally recovered. This technique is especially useful for the spent beams like in fig. 1.9 with a broad spectrum from 38 keV to higher than 130 keV.

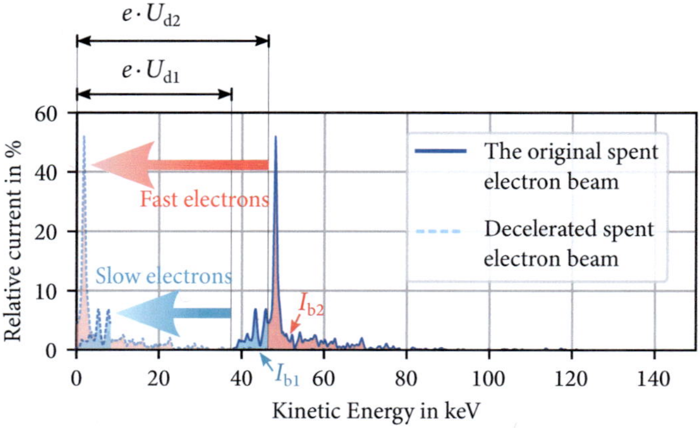

Fig. 1.12: Deceleration scheme for an ideal two-stage depressed collector

Analog to the ideal SDC deceleration scheme in fig. 1.10, fig. 1.12 shows the ideal deceleration scheme of the same spent electron beam using two depression voltages U_{d1} and U_{d2} with

$$U_{d2} > U_{d1} \,. \tag{1.29}$$

The low-energetic (slow) electrons with less than 45 keV energy form a beam current of I_{b1} (the cyan area), while the high-energetic electrons have a current of I_{b2} (the magenta area). Compared to the recovered electron energy of an SDC

$$P_{rec,SDC} = U_{d1} \cdot I_b = U_{d1} \cdot I_{b1} + U_{d1} \cdot I_{b2} \,, \tag{1.30}$$

the ideally recoverable energy with a two-stage MDC is

$$P_{rec,MDC} = U_{d1} \cdot I_{b1} + U_{d2} \cdot I_{b2} > P_{rec,SDC} \tag{1.31}$$

for the same beam energy spectrum. The reason that an MDC can have a higher efficiency than an SDC, is given in eqs. (1.29) to (1.31). If more than two stages can be applied, the collector efficiency can be even higher.

For example, let us consider a five-stage collector for the 170 GHz 1 MW gyrotron. Electrons in the spent electron beam have the initial kinetic energies varying from 38 keV to 130 keV. The depression voltage of the first stage should be fixed at 30 kV for the reason explained before. If the voltage increment ΔU_d between two adjacent stages is constant, the depression voltage will be

$$U_{\mathrm{d}n} = 30\,\mathrm{kV} + n \cdot \Delta U_\mathrm{d} \tag{1.32}$$

at the n'th stage. The theoretical collector efficiencies for $\Delta U_\mathrm{d} = 5, 10$ and $20\,\mathrm{kV}$ is shown in table 1.1. The collector efficiencies are higher than the one of an SDC (57.4 %). However, this is not the optimum. To achieve the optimal collector efficiency, a proper choice of the depression voltages is important.

Table 1.1: Theoretical collector efficiencies of a five-stage depressed collector using equidistant depression voltages for the realistic spent beam of the 170 GHz 1 MW gyrotron

ΔU_d (kV)	Depression voltages (kV)					η_col (%)
5	30.0	35.0	40.0	45.0	50.0	87.8
10	30.0	40.0	50.0	60.0	70.0	86.5
20	30.0	50.0	70.0	90.0	110.0	73.4

The maximal achievable efficiency and the optimal depression voltages of an MDC can be calculated for a given energy spectrum via a dynamic programming algorithm [51]. That is, in order to estimate the maximal efficiency and optimal depression voltages of an N-stage depressed collector, one has to solve the maximal efficiencies and the depression voltages for all possible U_d-combinations of $N-1$ stages; then, each MDC scenario of $N-1$ stages is calculated in the same way recursively based on the solutions of $N-2$ stages, so on and so forth until the case of an SDC, see appendix A.1. The estimated depression voltages will be the primitive input for the MDC principle designs.

The theoretical optimal depression voltages and the maximal collector efficiencies for the spent electron beam of the 170 GHz 1 MW gyrotron are given in table 1.2. The table shows that the optimal efficiency for a five-stage collector is much higher than using a set of non-optimal depression voltages (table 1.1). In case of a two-stage depressed collector, the sorted electron beam should be split at the kinetic energy slightly lower than the most probably electron energy. In this example the second depression voltage is 45.5 keV, while the highest peak in the spectrum is at approximately 50 keV, as visualized in fig. 1.12.

Table 1.2: Optimal depression voltages and collector efficiencies for the realistic spent beam of the 170 GHz 1 MW gyrotron

Stages	Depression voltages (kV)					η_{col} (%)
1	30.0					57.4
2	30.0	45.5				82.4
3	30.0	45.4	56.6			86.9
4	30.0	41.0	47.5	59.1		90.7
5	30.0	41.0	47.4	55.3	66.5	92.5

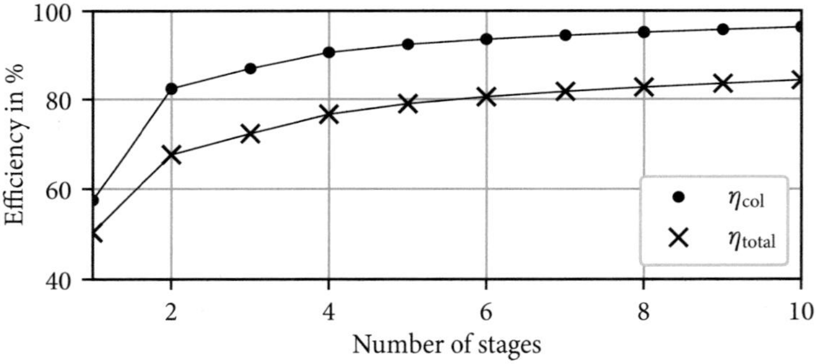

Fig. 1.13: Relation between efficiencies and number of stages assuming η_{int} = 35 % and η_{RF} = 90 %

Figure 1.13 shows the relation between the maximal achievable efficiencies and the number of collector stages. Both collector and gyrotron overall efficiencies increase with the number of stages. However, the contribution of an additional collector stage to the collector (and thus the overall) efficiency is no more significant when the number of collector stages is high. A two-stage depressed collector could theoretically raise the collector efficiency above 74 %, which means an overall efficiency higher than 60 %. A similar efficiency of a two-stage collector is estimated for the 140 GHz gyrotron for the W7–X stellarator and is also foreseen for the DEMO gyrotrons. However, all analyses here are based on the assumptions of an ideal energy sorting and an ideal recovering of the electron kinetic energy. Both assumptions are not achievable in reality. More than two stages might be required in the future. Therefore, the investigations on the MDC concepts should not exclude the possibility to upgrade the designs to more than two stages.

1.4.2 Multistage Depressed Collectors for Linear Beam Tubes

The idea of an MDC was reported even at the time when klystron was invented [52]. Today, MDCs can be found in many kinds of vacuum tubes, especially, they are applied in the Traveling Wave Tubes (TWT). Typical concepts and experimental behaviors of MDCs for these tubes are summarized in [53, 54]. Unlike the gyrotron beam shown in fig. 1.3, those devices have linear electron beams. That is, compared to the gyrotron hollow beam, a "filled" beam with a small outer radius r_b. There is only a very weak (in the order of 0.1 mT) longitudinal magnetic field affecting the beam electrons. The MDCs (for TWTs) have been systematically classified into 25 categories [54], as shown in fig. 1.14, where a circle means zero field, dashed and solid lines stand for the electric and magnetic field lines, respectively.

The most MDCs have symmetric fields, because the analysis and design of a symmetric MDC is usually simpler than an asymmetric one. All symmetric cases are enumerated in the top left corner of fig. 1.14. The effect of the magnetic field can be determined by universal beam-spread curves [55]. An MDC without or with a weak magnetic field can use an electrostatic lens to sort the beam electrons. Moreover, the potential and shape of electrodes can be analytically solved [56]. Figure 1.15 is an example of a "dispersive lens collector" calculated from equations. In such an MDC, the kinetic energy of the beam electrons cannot be totally recovered, because the space-charge force and the slant electrodes accelerate electrons radially. Ideally, such MDCs have the best performances if the stages have the smallest opening, through which all trajectories can just penetrate without impinging the far side of the electrodes, as shown by the upper trajectory in fig. 1.15, which touches the "0.6" electrode. On the other hand, if the electrons are collected on the far side, secondary electrons emitted from the far side can be pushed by the electric field back to the surface without causing a reduction of the efficiency.

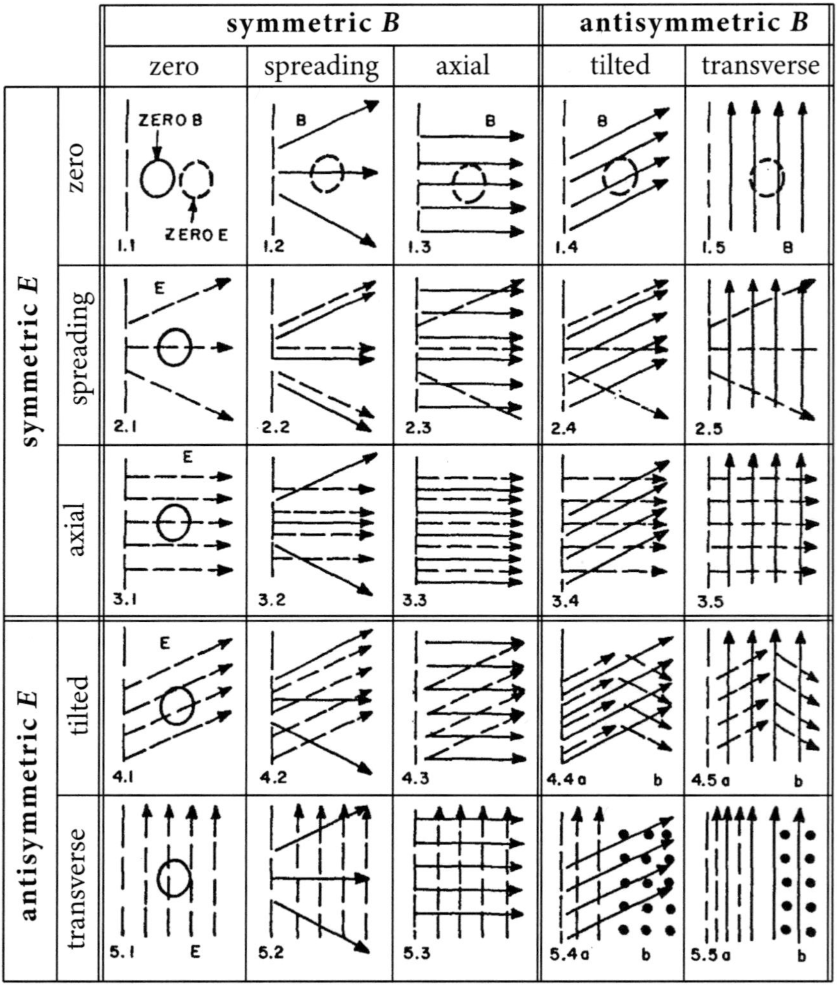

Fig. 1.14: Classification chart of TWT MDCs, reconstructed from [54]

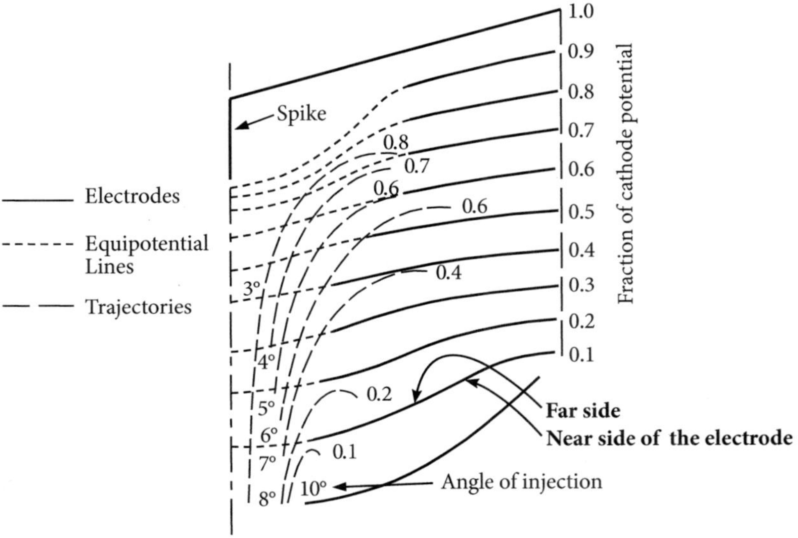

Fig. 1.15: Axisymmetric dispersive lens collector [53] (for TWT)

The asymmetric MDCs have also been studied in the history. The concept of an asymmetric MDC using a hyperbolic field is presented in [57]. Another example is the four-stage asymmetric MDC built for the Free Electron Laser (FEL) in [58], the scheme of which is shown in fig. 1.16a. The electrons will be accelerated and sorted by a transverse electric field to the "upper" electrodes in the figure. It has been noticed [59] that the introduction of a perpendicular magnetic field as shown in fig. 1.16b may significantly improve the collector performance for two reasons. First, the Lorentz force from the magnetic field compensates the transverse electric force, especially for the fast electrons. The electron trajectories are straighter than the ones without magnetic field in fig. 1.16a and the deceleration of the longitudinal velocities is more effective. Second, another effect which was not mentioned in [59] is the $\boldsymbol{E}\times\boldsymbol{B}$ drift. The drift also contributes to the sorting of electrons. However, in that study, the repulsion force between the electrons was so strong [60], that the introduction of an additional crossed magnetic field did not work as expected.

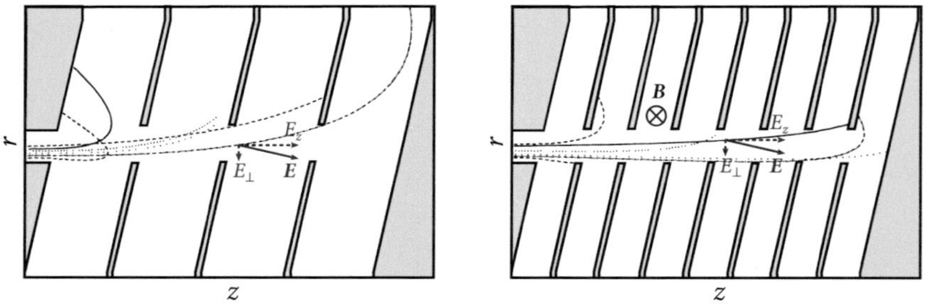

(a) only electric field (b) with a transverse magnetic field

Fig. 1.16: Trajectories in the asymmetric MDC scheme for FEL [59], the repulsion of space charges is ignored

The so-called Tilted Electric Field (TEF) Soft-Landing collector [61] was developed for TWTs in the 1970s. Figure 1.17 shows its structure. An additional axial magnetic field (28 mT in [61]) have to be maintained in the collector. Repulsion by the space charges, which was fatal in the concept of [60], is due to this axial magnetic field not an issue anymore. The linear beam passes through the off-center holes on each stage. During the traveling, electrons are sorted and steered by the drift towards electrode surface. Secondary electrons are driven by the same non-reciprocal drift and can be bounced maximally one stage backwards in the worst case, when the bombardment takes place on the near side of an electrode. The low-perveance experiment [61] of a five-stage collector has shown a very sharp sorting of electrons. According to [53], the collector efficiency in the experiment of a TWT was estimated to be 86 %. It was concluded in the study [54] over various MDC concepts that the TEF MDC is capable for an excellent efficiency, while it also avoids the issues with space charges and secondary electrons. The MDC study [54] claimed that "TEF collector has been selected as the best practical collector for future development". Nonetheless, investigations on TEF MDCs discontinued for nearly four decades. There could be multiple reasons that the TEF MDC was not commonly applied in TWTs:

- It requires an additional axial magnetic field [61], which is stronger than the usual axial guiding field. Providing this magnetic field increases the complexity of the system.

- Albeit the complexity, the TEF MDC is not more efficient than the best axisymmetric designs [53].
- Another challenge is the collector cooling in high power scenarios [61]. The cooling system seems to be more challenging than the axisymmetric MDC.

This concept is not directly usable in gyrotrons. The obvious reason is the difference of beam types.

There are some other concepts which rely on the crossed fields for the sorting of electrons. One of them is presented in [62]. It has the principle like a Wien filter [63]. The Lorentz force from a transverse magnetic field points to the opposite direction of the electric field. In this way, the fast electrons are steered mainly by the Lorentz force and can be collected on one side; while the electric force pushes the slow electrons to the opposite side. Electrons injected with moderate initial velocity experience the equilibrium between electric and magnetic forces, thus they can pass through this crossed-field region and are collected at the end stage with a moderate depression voltage. This concept is also not admissible for gyrotrons because of the magnetic confinement in a gyrotron collector.

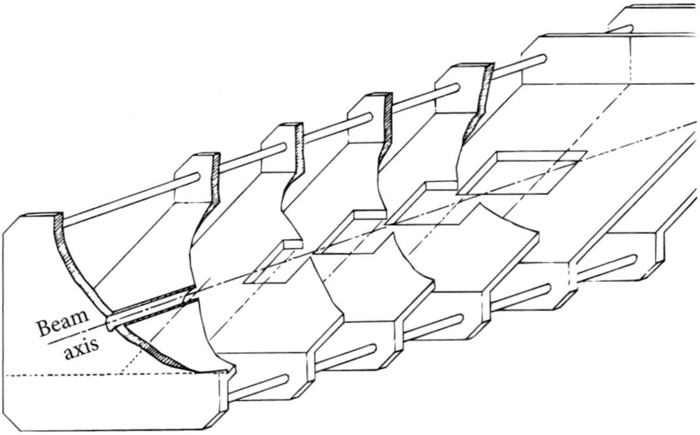

Fig. 1.17: Schematic of the tilted electric field collector [61]

1.4.3 State-of-the-Art for Gyrotron Multistage Depressed Collectors

Even though MDCs have been successfully applied in many other devices, there is yet no MDC developed for gyrotrons in use. Not only the construction and experiment of an MDC is more sophisticated than of an SDC, but it is also difficult to invent a proper concept, which can deal with the high magnetic field in gyrotrons. A fundamental harmonic gyrotron requires a strong (typically 6–10 T) magnetic field in the cavity according to eq. (1.3). After cavity, the magnetic field decays, however the flux density is still notable (typically 100 mT) at the entrance of the gyrotron collector.

Aside from the moderated magnetic field in the collector, the other challenge comes from the small-orbit electron beam. This small-orbit beam causes two effects:

1. As explained with fig. 1.6, it is hard to sort and separate the orbits radially.
2. To reduce (dilute) the magnetic flux density enclosed in the hollow beam, a large area of cross section is required. If the reduction of the magnetic field should be adiabatic, the transition needs a long distance. The large area and long length mean a huge volume for the collector.

The well-developed MDC technologies for klystrons and (TWTs) [53, 54] are not directly applicable because of the moderated magnetic field. There was an MDC designed and built for an X-band gyrotron Backward-Wave-Oscillator (BWO) [64, 65]. However, the magnetic flux enclosed in the electron beam of a fusion gyrotron is approximately 30 times higher than the one in that gyro-BWO. The size of that MDC-type will also be unacceptable in a fusion gyrotron.

There are several theoretical proposals of MDC concepts for the fusion gyrotrons. They can be classified into two categories:

1. the axisymmetric concept,
2. the non-axisymmetric concept, which uses $\boldsymbol{E}\times\boldsymbol{B}$ drift.

The state-of-the-art of both proposals are presented in the following paragraphs. Advances beyond the state-of-the-art in both concepts will be presented in chapters 2 and 3, respectively.

Axisymmetric Proposals

Axisymmetric MDC involves only axisymmetric electric and magnetic fields, which implies that all coils are centered around the collector axis. It is possible to describe such an MDC model and its fields in a two-dimensional system. Therefore, the geometrical definition and simulation of such an axisymmetric MDC requires only a relatively small effort.

As mentioned before, there were some investigations on the MDC for large-orbit gyrodevices [6, 64–66]. However, those proposals should operate in a weak (much lower than 10mT) magnetic field. For example, the scheme in [6] has 6 mT field in the collector, compared to a fusion gyrotron, which has a magnetic field up to 100 mT in the collector. In the low magnetic field, the energy sorting of the large-orbit electron beam is performed partly by the applied electrostatic field and partly by the repulsion force of space charges.

In small-orbit gyrotrons the situation is different. Electrons in a small-orbit gyrotron SDC rotate within a Larmor radius of less than 3 mm, while their guiding centers can be further than 200 mm (in the collector) away from the center axis. The existing axisymmetric MDC proposals for small-orbit gyrotrons require a magnetic field lower than 10 mT. The demagnetization from several Tesla's in the cavity down to the magnitude less than 10 mT in the collector is preferred to be adiabatic, since an adiabatic transition can convert $v_\perp$ to $v_\parallel$, according to eq. (1.20). If electrons in such a low magnetic field have mainly $v_\parallel$, they can be sorted and collected like in the MDCs for other tubes.

The disadvantage of the adiabatic demagnetization is a huge collector size, as discussed on page 34. In order to limit the collector length, the demagnetization should be slightly non-adiabatic [7–10]. To reduce the radius, coils or pole shoes like in fig. 1.18 are inserted coaxially inside the electron beam [10, 11], so that the magnetic flux enclosed by the annular electron beam is condensed within the inner coils or pole shoes around the axis.

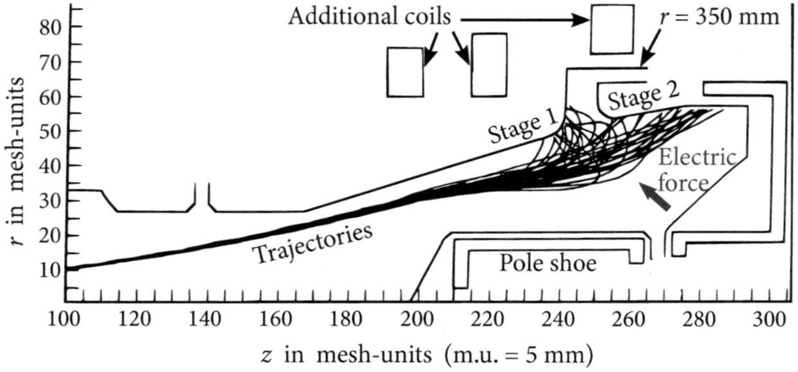

Fig. 1.18: The first sub-type of axisymmetric MDCs for a 110 GHz, 1 MW gyrotron [10]

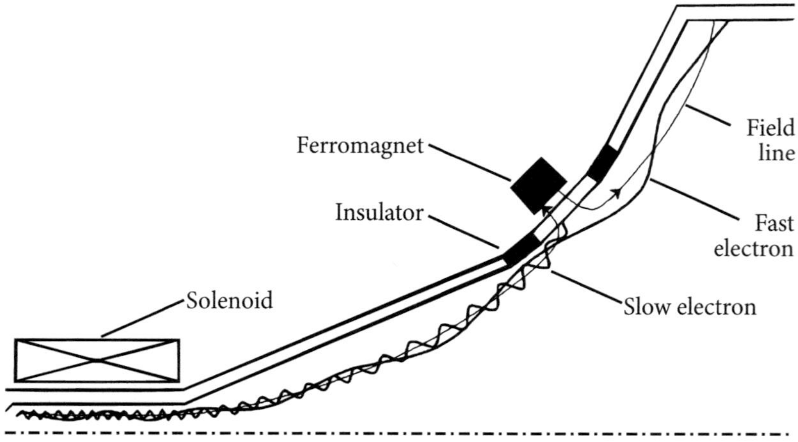

Fig. 1.19: The second sub-type of axisymmetric MDCs using a so-called "magnetic separator" [12]

There are two sub-types of axisymmetric gyrotron MDCs in literature. In the first sub-type, the electrostatic force dominates the sorting of electron trajectories [10]. Accordingly, the magnetic field should be weak (less than 10 mT), and the demagnetization is relatively adiabatic. Figure 1.18 shows the simulation model of a gyrotron MDC of this kind. This two-stage collector for a 110 GHz (B_{max} = 4.3 T) gyrotron [67] has a maximum radius of approximately 350 mm.

According to Fig. 4 of its preceding publication [9], the magnetic flux density at the place where the beam starts to diverge ($z \approx 220$ mesh units) is approximately

$$B = \frac{(24000 - 18000) \cdot 0.1\,\text{mT} \cdot \text{cm}^2}{\pi\left[\left(37\,\text{m.u.} \times 0.5\frac{\text{cm}}{\text{m.u.}}\right)^2 - \left(25\,\text{m.u.} \times 0.5\frac{\text{cm}}{\text{m.u.}}\right)^2\right]} \approx 1\,\text{mT}\,.$$

In such a weak magnetic field, the electric force marked in fig. 1.18 pushes the slow electrons towards the first stage, while the trajectories of fast electrons are less affected.

The maximal radius of this design is 350 mm. It already meets the constraint of the expected MDC radius. Up-scaling the gyrotron frequency by 1.55 times (from 110 GHz to 170 GHz), the magnetic field strength and flux also increase approximately by 55 %, hence, an even larger collector radius will be required. The difference between the magnetic field in the cavity and in the collector also becomes larger when the concept is up-scaled. Therefore, the adiabatic demagnetization will require an even longer transition length. In case of a future DEMO gyrotron operating at 204 GHz, it will be impossible to up-scale this concept considering a reasonable collector size.

The second sub-type of axisymmetric gyrotron MDC is depicted in fig. 1.19. A ferromagnetic object affects the magnetic field locally. There, slow electrons follow the local magnetic field lines strictly such that they are collected at the stage beneath the ferromagnet, while the trajectories of fast electrons are less affected by the local alteration of magnetic field, and they continue to move to the next stage. In the literature [12], this scheme is called a "magnetic separator". However, it should follow Busch's theorem, thus the electron orbits cannot be really separated, if the slow electrons could not be captured in time. Moreover, the creation of such a precise magnetic field is tricky. There is no example of any principle design proposed in the literature.

Non-Axisymmetric Proposals Using *E*×*B* Drift

The idea of a gyrotron MDC, which uses the ***E***×***B*** drift to sort electrons, was proposed for the first time in [68]. There was no other discussion about this kind of gyrotron MDC until the presented work.

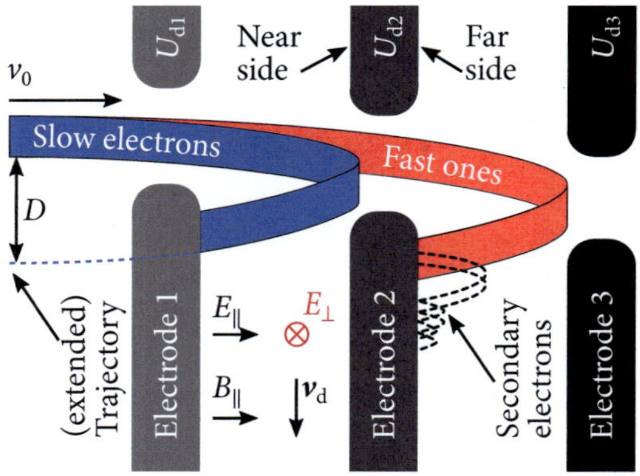

Fig. 1.20: Scheme of an $\boldsymbol{E}\times\boldsymbol{B}$ MDC using $E_{\perp}$

The principle of a gyrotron $\boldsymbol{E}\times\boldsymbol{B}$ MDC is shown in fig. 1.20. There, the electron beam is confined and guided by a longitudinal magnetic field $B_{||}$. Ideally, the axisymmetric magnetic field in the collector would be homogeneous and would never decay. There is a longitudinal component of electric field $E_{||}$, which decelerates the electrons. The electrons have longitudinal velocities v_0 parallel to the axis. Due to $E_{||}$, the further electrons longitudinally reach, the higher is the recovered kinetic energy. Aside from the deceleration field, there is another component of electric field $E_{\perp}$ perpendicular to the plane. This is the main difference to the axisymmetric concepts before, where the electric field was only in the projected (r, z)-plane. The perpendicular component of electric field causes an $\boldsymbol{E}\times\boldsymbol{B}$ drift given by eq. (1.15), which is here equivalent to

$$\boldsymbol{v}_{\mathrm{d}} = \frac{E_{\perp}}{B_{||}} \,. \tag{1.33}$$

All electrons have the same drift velocity, which is independent of mass, charge and (even the direction of) initial velocity; however, the exposure time of each electron in the drift field is related to its initial longitudinal velocity v_0 (which is close to its total velocity when α_{p} is low). When a slow and a fast electron move forward and travel the same longitudinal distance, the slow electron has been exposed in the drift for a longer time than the fast one. Accordingly, the slow electron had a longer drift distance, which is sufficient to let the electron be collected at one of the first stages near the entrance. The next stages decelerate the originally fast electrons further, so that they have sufficient time to drift to one of the stages with a higher depression voltage. In this way, electrons are sorted by their initial (longitudinal) velocities, which are related to their initial kinetic energies.

Secondary electrons emitted from the far side of an electrode suffer from the same drift velocity as the primary ones. The drift velocity is independent of the emission angle. At the same time, the deceleration field $E_{\|}$ pushes secondary electrons back to the electrode surface where they were emitted. Those secondary electrons emitted from the near side of electrode 2, 3, 4 ... can bounce maximally only one stage back in the existence of the drift. In the worst case, they will be collected on the far side of the previous electrode. This would decrease the collector efficiency, but it prevents secondary electrons from bouncing more than one stage backwards.

In fig. 1.20, the key is to create the electric field component $E_{\perp}$ perpendicular to the initial direction of electron guiding centers. However, Faraday's law

$$\boldsymbol{\nabla} \times \boldsymbol{E} = -\mu \frac{\partial \boldsymbol{H}}{\partial t} \tag{1.34}$$

prevents the electric field from building a closed loop in any azimuthal cross section; otherwise the magnetic flux should quickly vary with time, which is not possible in a gyrotron collector. Therefore, the radial drift should change its sign along a closed integral path of electric field. Hence, the axisymmetry is broken, if the integral path is on an azimuth circle. The break of axisymmetric becomes the first challenge for the conceptual design of such an MDC.

Figure 1.21 shows the first idea of the gyrotron $\boldsymbol{E}\times\boldsymbol{B}$ collector originally from [68]. There, the electron beam moves between two cylindrical layers of electrodes. These electrodes are supposed to be helical (cooling) pipes. Adjacent pipes have different electric potentials and the depression voltage changes gradually from one pipe to another. The torsion of the pipe around the center axis creates a transverse as well as a longitudinal electric field component. At any transverse cross section, there is a clockwise rotating azimuthal electric field $E_{\theta} = E_{\perp}$ on half of the circle, and a counter-clockwise electric field $E_{\theta} = -E_{\perp}$ on the other half. There is a uniform longitudinal magnetic field. The azimuthal electric field component E_θ and the longitudinal magnetic field produce a radial inward drift $\boldsymbol{v}_{\mathrm{d,i}}$ and an outward drift $\boldsymbol{v}_{\mathrm{d,o}}$, depending on the sign of E_θ, i.e. on which half side of the cross section the electron is located. If an infinite number of stages (electrodes) is assumed, simulations using the electron-optics code ARIADNE [69] show a collector efficiency up to $\eta_{\mathrm{col}} = 92\,\%$.

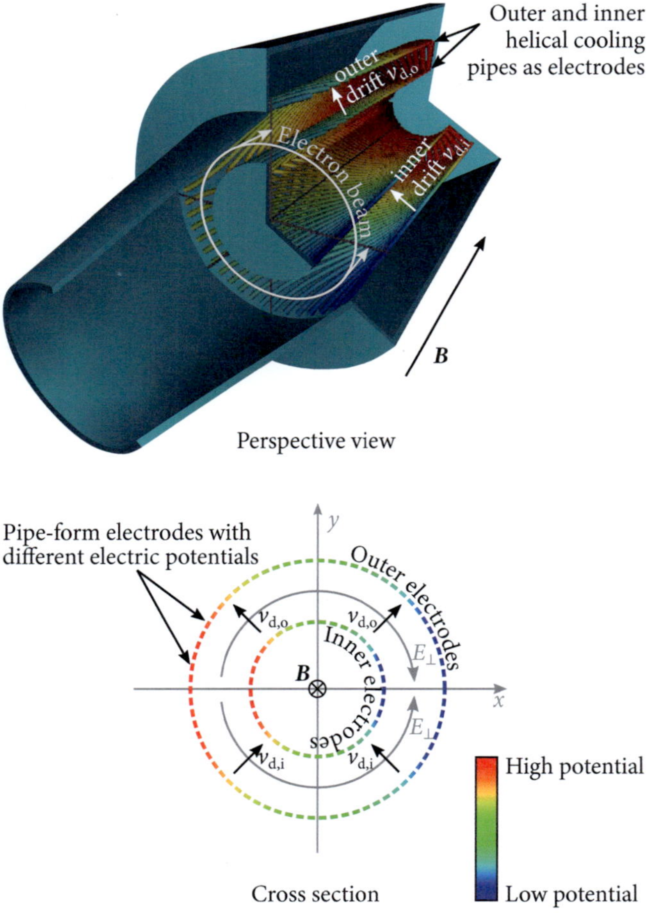

Fig. 1.21: The original idea of gyrotron $\boldsymbol{E}\times\boldsymbol{B}$ MDC with an infinite number of stages [68]

1.5 Critical Parameters and Constraints of the Multistage Depressed Collector Designs in this Work

The goal is to find the most proper MDC concept for the future 2 MW coaxial DEMO gyrotron operating at frequencies of 170 GHz and 204 GHz. As the DEMO gyrotron is in the early stage of development [70], the 170 GHz 1 MW gyrotron with conventional hollow cavity will be taken as the reference gyrotron for the MDC designs. Those MDC designs will take the size and depression voltage of the reference SDC (fig. 1.11) as the starting design parameters, because 170 GHz is also a target frequency of the DEMO gyrotron. Two fundamental-harmonic gyrotrons having the same frequency means that their maximal magnetic fields are similar. The difficulty of energy sorting in the MDC is related to the maximal magnetic field at the cavity for a given collector size. Therefore, the final MDC design should be able to operate with the 170 GHz DEMO gyrotron and should be able to up-scale to the 204 GHz operation point. Moreover, the DEMO gyrotron will deliver 2 MW RF power, whereas the reference gyrotron only has 1 MW. Both will have a similar acceleration voltage. A first approximation for the collector design is to double the beam current. Hence, the power of the spent electron beam is also approximately doubled in the case of a 2 MW gyrotron.

Several MDC conceptual designs for the reference gyrotron will be proposed and investigated in chapters 2 and 3. These designs will be assessed mainly from four aspects:

1. Achievable collector size and mechanical complexity
2. Expected collector efficiency
3. Unwanted back-stream current
4. Maximal expected power load density of the critical parts

Secondary electrons cause two effects: the reduction of the collector efficiency and a back-stream current. They are considered separately in the second and third points, respectively.

Achievable Collector Size and Mechanical Complexity

Since the most significant difference of a gyrotron collector to the collectors for other tubes is the magnetic field, reducing the magnetic field might be a solution. In order to reduce the field in the collector, the conventional collector proposals (e.g. [6, 8, 11]) require a large collector size, especially for the high frequency fusion gyrotrons. Hence, the collector size is the first parameter to be considered.

After the gyrotron is assembled, the whole tube should be baked out in an oven for a good vacuum condition inside. The gyrotron, including the collector, should be able to fit into the oven. At KIT, the oven is more than 4 m long and has the maximal width of 0.8 m. Derived from the size of the oven, the collector should not be longer than 2 m (doable). The inner radius of the collector can maximal be around 300 to 350 mm (challenging). The reference SDC is 1.3 m long, with an inner radius of 225 mm. However, that SDC is just a simple hollow cylinder, while an MDC will have more complicated structure (e.g. isolated electrodes). To contain the additional components of the structure, an MDC will be wider than an SDC. Another reason for a larger size of an MDC is that a specific magnetic field should be applied, whereas the gyrotron magnetic field cannot directly be used and is usually too strong for the multistage collection. A large cross section is needed to dilute the magnetic field from the gyrotron coil. Hence, the radius of an MDC is more critical than the length.

The complexity of an MDC consists of the complexity of electrodes and the one of coils. An MDC involves multiple stages, the shapes of electrodes at each stage and their topology are preferred to be simple. Additional coils are also necessary in a gyrotron MDC to shape and tune the collector magnetic field. These coils should not be super conducting. The MDC design should also take the number and complexity of these coils into account.

Expected Collector Efficiency

An MDC should have a collector efficiency higher than 74 % in order to raise the overall gyrotron efficiency above 60 %. The efficiencies of the MDC designs with the same stage number and similar size will be compared for the same spent beam.

The precision of the energy sorting will be checked with mono-energetic beams in the simulations of the representative designs. In addition, the sensitivities of collector efficiency against the space charge effect (via varying the beam current), the perturbation magnetic field and the electron beam displacement will be assessed for the representative designs.

An MDC might have more than two stages in order to achieve an even higher efficiency. The possibility to extend the number of depression stages will also be assessed for the MDC concepts.

Secondary electrons may cause inter-stage currents, which work like a resistor (or even a short-circuit) connecting the stages. These currents reduce the MDC efficiency. To check this effect, the efficiencies of collector designs with and without considerations of secondary electrons will be compared.

Unwanted Back-Stream Current

In case there is a beam current streaming back to the cavity, the energy carried by the back-streamed (reflected) electrons are considered to be lost in the calculation of collector efficiency. In addition, the back-streamed electron current might affect the interaction, which would not only cause the reduction of the interaction efficiency and the RF power, but could also exceed the capability of the power supply, if too much of this current is collected by the gyrotron body (the current body power supply at KIT only allows 200 mA of body current). Therefore, the back stream of beam current should be suppressed.

There are two sources for the back-streamed beam current. A part of the primary beam electrons may not be appropriately decelerated, such that they could be reflected by the depression voltage from the electrodes or from the space charges. The other source is the secondary electrons emitted at each stage. The decelerating electric field for the primary electrons becomes an accelerating electric field for the secondary electrons towards the entrance of the collector. If these secondary electrons cannot be intercepted before they leave the collector, they will be a part of the back-stream current. The back-stream current of the MDC designs in chapters 2 and 3 will be checked.

Maximal Expected Power Load Density on the Critical Parts

Single-stage depressed collectors of CW gyrotrons should have a mechanism to sweep the impacting position of the high-energy spent electron beam, such that the power load density at the collector is averaged during the operation time. In the reference SDC, the maximum at any phase of the sweeping is higher than 2.5 kW/cm^2. When the electrons impact the electrodes, their energy raises the local temperature on electrodes, thus the material deforms. With the beam sweeping, there is a periodic thermal deformation which causes stresses of the material. The stress is associated with the material fatigue, which limits the collector lifetime. The current limitation of the temporally averaged load density is 500 W/cm^2 given by THALES [50].

If the load is static, there is no periodic stress in the material caused by the sweeping and the acceptable static power load density might be higher than the swept average value. There is no specification considered in this work for such static loads, but some studies can be found in the literature. For example, the nominal MDC load density in [6] is 1 kW/cm^2. The thermal analysis in [23] has shown that a power density of 900 W/cm^2 at the second stage of that gyrotron MDC is acceptable for a 500 ms operation.

The maximal expected power load of the MDC proposals in chapters 2 and 3 will be given for the 170 GHz 1 MW gyrotron electron beam. For a DEMO 170 GHz 2 MW gyrotron, the load density is expected to be doubled. In order to reduce the power load density, a local sweeping mechanism might also be considered in the design of gyrotron MDCs.

1.6 Collector Simulation Techniques and Codes Used in this Work

The geometry and field of a gyrotron SDC [71, 72] as well as the conventional gyrotron MDCs [10–12, 64, 73] are axisymmetric. Two-dimensional axisymmetric particle-beam-optics codes for the Magnetron Injection Guns (MIG) can also be applied for symmetric collectors. At KIT, ARIADNE [69] and ESRAY [38] are the beam optics codes for the simulations of axisymmetric single or multi-stage collectors.

Typically, the codes simulate the collector system with the so-called trajectory TRacKing (TRK) method. Structured high-order meshes are generated via domain mapping for an axisymmetric collector geometry. The static magnetic field is calculated by methods derived from the Biot-Savart integration, which can be highly parallelized. ARIADNE and ESRAY solve firstly the electrostatic problem i.e. the Poisson equation

$$\nabla^2 \varphi = -\frac{\rho_\mathrm{q}}{\varepsilon} \tag{1.35}$$

by finite element and finite difference methods, respectively. φ is the electric potential, ρ_q is the charge density and ε is the permittivity. Then the electron beam is tracked in the electrostatic and magnetostatic fields. As the simulation of each single physical electron is not possible due to the number of involved electrons, the electron beam is represented by a set of macro electrons. A macro electron presents a packet of numerous physical electrons together and it is driven by the Lorentz force. The beam electrons are not directly aware of the existence of each other, when they are driven by the Lorentz Force. The motion of electrons is only affected by the static electric and magnetic field from the mesh grids. Therefore, the tracking of each particle trajectory can be highly parallelized. Once the trajectories of all beam electrons have been tracked, the Poisson equation is recalculated. The recalculation takes the space charges within the trajectories from the last tracking iteration into account. Depending on the simulation settings, the magnetic field may also be updated. (Nevertheless, the induced magnetic field from the gyrotron electron beam is negligible compared to the existing magnetic field in a gyrotron.) In the next step, the electron trajectories are tracked once again. So far and so forth, until the system eventually reaches a stationary state after 10–20 iterations.

The Particle-In-Cell method (PIC) [74] is used to crosscheck the results from the TRK method. The PIC is a transient method, where the currents and charges of the particles are aggregated in cells to save the computational complexity caused by the direct interaction of plenty macro electrons. The PIC verification of a gyrotron collector takes weeks of simulation time on a today's PC to evolve to a stationary state. For a typical gyrotron collector, the stationary state occurs after several hundreds of nanoseconds physical time, while the pulse lengths in the short-pulse experiments [24, 26] are usually in millisecond range.

There are several methods to describe geometries in a three-dimensional space, for example, using explicit or implicit (in-) equations, voxels, Constructive Solid Geometry (CSG), surface meshes, Boundary REPresentation (BREP) in particular using Non-Uniform Rational B-Splines (NURBS), etc. It is non-trivial to implement a robust geometry core for the simulation tools even only including basic geometric features, such as Boolean operation of objects, surface interaction, loft and rail operation, etc. Fortunately, handling such three-dimensional geometries was not necessary in axisymmetric gyrotron simulations. However, if one would like to investigate the collector types which break the axisymmetry for various reasons, the full three-dimensional description of geometries and mesh generation are unavoidable. CST Studio Suite® is used in this work for the simulation of asymmetric MDCs. Both TRK and PIC are implemented in CST, where the PIC involves the full electromagnetism, rather than electrostatic. Although CST supports unstructured tetrahedral meshes, hexahedral meshes are mainly used in the MDC simulations due to the lack of the possibility for field importation at the time of writing. As the magnetostatic solver in CST needs some efforts and notable computational resources to deliver accurate results, in the most cases, the collector magnetic field is calculated via the Biot-Savart law externally, see appendix A.2. More details of the three-dimensional simulations will be given in chapter 3.

The computational resource required by a gyrotron collector simulation often exceeds the capability of a today's PC. The high requirement of resources is mainly due to the large geometry and large number of (macro) electrons:

- The size of an SDC is large, due to the high magnetic flux and the dissipation of the high power. For example, the 170 GHz 1 MW gyrotron has 2.4–2.6 MW power in the spent electron beam. Its SDC has an inner radius of 225 mm and a length of 1.33 m. For a future 2 MW DEMO gyrotron, the spent beam power will be 4.7 MW (estimated from [75]). The collector will also be larger in that case. Huge geometry requires numerous mesh cells, especially in the three-dimensional case. A three-dimensional geometric and field representation is necessary for the $\boldsymbol{E}\times\boldsymbol{B}$ drift concept (see section 1.4.3) and for the investigations of the asymmetric effects on the symmetric concept (e.g. transverse sweeping of the electron beam, beam misalignment, perturbation field, etc.) On the other hand, an MDC involve more complicated geometry, to resolve that, even more mesh cells are needed. The large number of mesh cells slows down the simulation.

- The injected electron beam has to be sampled with plenty macro electrons in order to have a good statistic representation. What makes the situation even worse is that the electrons should be decelerated in the collector, such that each electron stays in the simulation region for a long time. In the PIC simulations, the motions of all electrons have to be evaluated at each time step; therefore, the effort for the calculation of these numerous electrons at each time step is huge. Additionally, the emission of secondary electrons introduces even more electrons into the system. The extremely large number of primary and secondary electrons needs a large amount of computational resources, both in time and memory, for the updating and storage of the particle status.

2 Proposal for an Advanced Axisymmetric Collector Concept Allowing a High Magnetic Flux within the Electron Beam

2.1 Advances to the State-of-the-Art in Brief

The first sub-type of the axisymmetric MDC concepts relies on two features [10]:

> (1) demagnification so as to transform energy from gyrating motion to longitudinal motion as far as possible and (2) to make the force from the electrostatic field dominate that from the magnetic field for effective energy sorting.

This kind of MDC is suitable for a gyrotron with an electron beam enclosing a relatively low magnetic flux (low B or small r_{gc}). Even though, a large collector size is necessary. The MDC design proposal in fig. 1.18 for a 110 GHz gyrotron [67] with $B_{max} = 4.3$ T in the cavity already requires a radius of 350 mm. This design is not practical for the 170 GHz or 204 GHz DEMO gyrotron. First, the expansion of the beam radius in order to dilute the magnetic flux will be significant (related to ψ). Second, the transition (related to B_{max}) from the high magnetic field at the entrance of the collector to a low field (approximately 1 mT) in the collector will take a longer distance than in the case of the 110 GHz gyrotron.

To overcome the expansion of collector size by the up-scaling to a high-flux gyrotron, a new sub-type of axisymmetric MDC is introduced in this chapter. A significantly stronger non-adiabatic magnetic field transition is necessary, because only with that, the size of an axisymmetric MDC can be controlled in the acceptable range. This new kind of axisymmetric MDC breaks both rules quoted above, that means:

1. The transition of magnetic field is strongly non-adiabatic.
2. The sorting of electron beam is dominated by the magnetic field instead of the electric field. The phase of gyrating motion is used to sort electrons.

The details of this advanced axisymmetric MDC proposal will be presented in the next sections.

2.2 New Principle for the Sorting of Electron Energies in an Axisymmetric System

With a certain type of non-adiabatic transition, the gyrations of beam electrons can be spatially modulated — the non-adiabatic transition "amplifies" the transverse motion of electrons and brings the trajectories oscillating radially. The kinetic energies of electrons are modulated on the amplitude (radius) and period length of these oscillations. This spatial modulation is demonstrated in fig. 2.1. For a clear demonstration, an ideal (but still valid for the Maxwell equations) analytical magnetic field is used. The idea behind this analytical magnetic field is to create a transition from a homogeneous field $\boldsymbol{B}_1 = \hat{z}B_1$ at $z < -z_0$, to another homogeneous field $\boldsymbol{B}_2 = \hat{z}B_2$ at $z > z_0$, as shown in eq. (2.1). A brief explanation for eq. (2.1) is given in appendix A.3.

$$B_z(z) = \begin{cases} B_1 & z < -z_0 \\ B_1 + \left(\frac{1}{5}\frac{1}{{z_0}^4}(z^5 + {z_0}^5) - \frac{2}{3}\frac{1}{{z_0}^2}(z^3 + {z_0}^3) + (z + z_0)\right)\zeta & |z| \leq z_0 \\ B_2 & z > z_0 \end{cases} \tag{2.1a}$$

$$B_r(z,r) = -\frac{1}{2}\zeta r \cdot \begin{cases} \left[1 - \left(\frac{z}{z_0}\right)^2\right]^2 & |z| \leq z_0 \\ 0 & \text{otherwise} \end{cases} \tag{2.1b}$$

where

$$\zeta = \frac{15}{16}(B_2 - B_1) \cdot \frac{1}{z_0} \quad . \tag{2.1c}$$

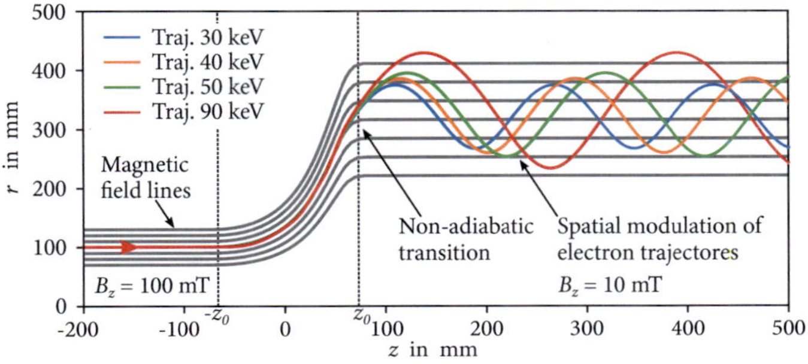

(a) Spatial modulation of electron trajectories

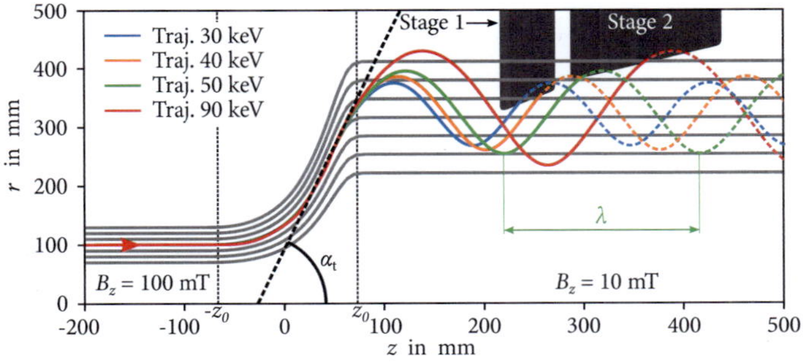

(b) Place the electrodes at proper positions

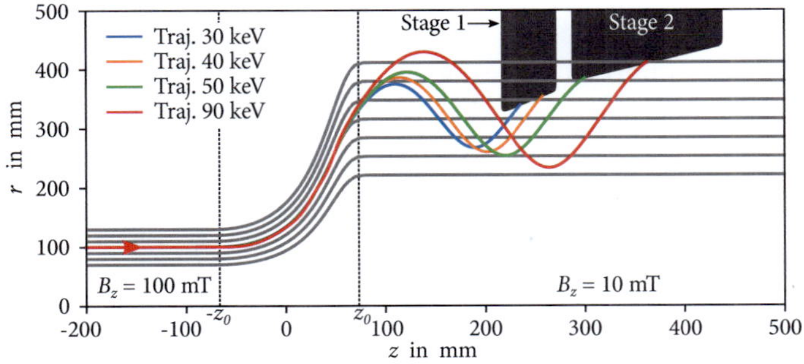

(c) An ideal two-stage depressed collector

Fig. 2.1: Principle of the advanced axisymmetric MDC based on the spatial modulation of electron trajectories

In this particular example, following parameters are considered:

$$B_1 = 100\,\mathrm{mT}\,, \quad B_2 = 10\,\mathrm{mT}\,, \quad \text{and} \quad z_0 = 80\,\mathrm{mm}\,.$$

For a simplified mathematical modeling, the transition of the magnetic field is the only object to be examined. There is yet no electric force applied. This model will already be capable of explaining several properties of such an MDC system, those can also be observed in the simulations considering electric forces and non-ideal fields.

Electrons are injected from the left-hand side ($z < -z_0$) to the system. Initially, the electrons only have longitudinal components of velocities for a clear demonstration. Of course, the modulation still exists even if the injected electrons have initial transverse velocities, as will be proven in section 2.3. Instead of trying to "separate" trajectories like fig. 1.19, this principle manages the local phases of the cyclotron motions.

After an electron passes through the transition region ($z > 80\,\mathrm{mm}$ in fig. 2.1), two parameters are modulated on the trajectories:

1. the maximally reachable radius r,
2. the longitudinal period λ of the cyclotron motion.

Both parameters increase with the initial kinetic energy of the electron. If the electron beam has a spread of initial transverse velocities, the tendency of this observation is still valid, however, an error interval should be considered (see the analysis in section 2.3). In this way, the phases of the electron cyclotron motions are sorted at their second half period after the non-adiabatic transition. In fig. 2.1 this happens between $220\,\mathrm{mm} < z < 400\,\mathrm{mm}$. With properly placed electrodes it is possible to design an MDC based on this effect.

2.3 Analysis of the Electron Energy Sorting Model

2.3.1 Introduction of the Theoretical Model

A theoretical model can be abstracted from the demonstration in fig. 2.1. As no electric field will be considered in this model, the kinetic energy of a particle does not change, and the relativity is ignored. The formulation begins at the right-hand-side boundary of the transition ($z = z_0$), where the electrons are leaving the non-adiabatic transition and traveling into a homogeneous field $\boldsymbol{B} = \boldsymbol{z}B_z$ with velocity components V_z, V_r, V_θ. These components are capitalized, because they are treated as the initial states, which might have random spreads even for the electrons with the same kinetic energy. V_z and V_r are positive real numbers. The lowercase variants v_z, v_r, v_θ are the components of the time-dependent velocity, which are evolved from the initial states. All electrons are approximately at the initial radius R_0 when they leave the transition region. For example, in fig. 2.1 $R_0 = 350\,\text{mm}$. (This is just an example using some arbitrary parameters to demonstrate the non-adiabatic transition. A size-optimized MDC design may have a different set of parameters.)

The cyclotron motion of an electron is depicted in fig. 2.2. The y-axis of the system is aligned to the vector $\boldsymbol{R}_0$, which means

$$\boldsymbol{R}_0 = \begin{pmatrix} 0 \\ R_0 \end{pmatrix} . \tag{2.2}$$

The initial transverse velocity $\boldsymbol{V}_\perp$ consists of the azimuthal and radial components, where the initial radial speed is aligned to $\boldsymbol{R}_0$ (therefore, also to the y-axis):

$$\boldsymbol{V}_\perp = \begin{pmatrix} V_\theta \\ V_r \end{pmatrix} . \tag{2.3}$$

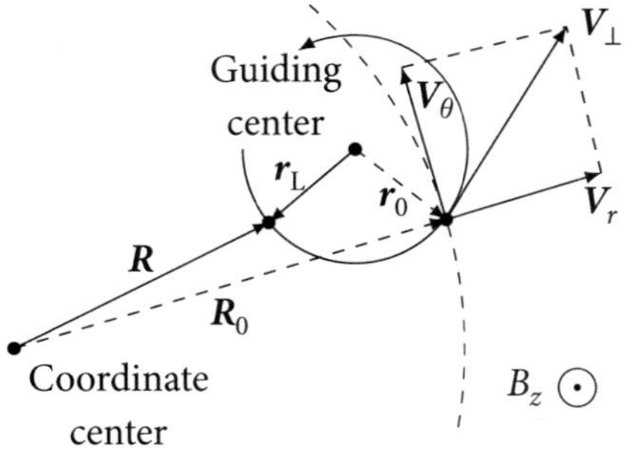

Fig. 2.2: Projection of the cyclotron motion in a homogeneous magnetic field

The cyclotron frequency ω_c does not change its sign in the process, therefore, it is free to let ω_c be positive. The cyclotron frequency is

$$\omega_c = \frac{|\boldsymbol{V}_\perp|}{|\boldsymbol{r}_0|}\,, \tag{2.4}$$

where $\boldsymbol{r}_0$ is the vector of the initial Larmor radius. As shown in fig. 2.2, $\boldsymbol{r}_0$ is perpendicular to the initial transverse velocity $\boldsymbol{V}_\perp$:

$$\boldsymbol{r}_0 = G(-90°)\frac{\boldsymbol{V}_\perp}{\omega_c}\,, \tag{2.5}$$

where the matrix $G(\phi)$ is the affine rotation by angle ϕ, which is in a two-dimensional Cartesian coordinate

$$G(\phi) = \begin{pmatrix} \cos\phi & -\sin\phi \\ \sin\phi & \cos\phi \end{pmatrix}. \tag{2.6}$$

These are the initial invariables for the trajectory of an individual electron.

For an electron gyrating in the homogeneous magnetic field, its Larmor radius does not vary, however, the direction of the Larmor radius vector $\boldsymbol{r}_L$ rotates with the cyclotron frequency ω_c. The initial state is $\boldsymbol{r}_L = \boldsymbol{r}_0$ at $t = 0$, then, this vector is rotated by an angle of $\omega_c t$ at time t:

$$\boldsymbol{r}_L = G(\omega_c t)\,\boldsymbol{r}_0\,. \tag{2.7}$$

As an outcome of this formulation, the global radius $\boldsymbol{R}$ of an electron, which is graphically presented in fig. 2.2, yields

$$\begin{aligned}
\boldsymbol{R} &= \boldsymbol{R}_0 - \boldsymbol{r}_0 + \boldsymbol{r}_L \\
&= \begin{pmatrix} 0 \\ R_0 \end{pmatrix} + \begin{pmatrix} \cos\omega_c t - 1 & -\sin\omega_c t \\ \sin\omega_c t & \cos\omega_c t - 1 \end{pmatrix} \boldsymbol{r}_0 \\
&= \begin{pmatrix} 0 \\ R_0 \end{pmatrix} + \frac{1}{\omega_c} \begin{pmatrix} \cos\omega_c t - 1 & -\sin\omega_c t \\ \sin\omega_c t & \cos\omega_c t - 1 \end{pmatrix} \begin{pmatrix} 0 & 1 \\ -1 & 0 \end{pmatrix} \boldsymbol{V}_\perp\,,
\end{aligned}$$

where those two transformations before $\boldsymbol{V}_\perp$ can be merged:

$$\boldsymbol{R} = \begin{pmatrix} 0 \\ R_0 \end{pmatrix} + \frac{1}{\omega_c} \begin{pmatrix} \sin\omega_c t & \cos\omega_c t - 1 \\ 1 - \cos\omega_c t & \sin\omega_c t \end{pmatrix} \begin{pmatrix} V_\theta \\ V_r \end{pmatrix}. \tag{2.8}$$

As a consequence of the Cauchy-Schwarz inequality, the maximal reachable radius is

$$\begin{aligned} R_{\mathrm{max}} &= |\boldsymbol{R}_0 - \boldsymbol{r}_0| + |\boldsymbol{r}_{\mathrm{L}}| \\ &= \left(\sqrt{|\boldsymbol{V}_\perp|^2 + 2\omega_{\mathrm{c}} R_0 V_\theta + \omega_{\mathrm{c}}{}^2 R_0{}^2} + |\boldsymbol{V}_\perp|\right) / |\omega_{\mathrm{c}}| \,. \end{aligned} \tag{2.9}$$

Corresponding to fig. 2.2, the term $|\boldsymbol{R}_0 - \boldsymbol{r}_0|$ in eq. (2.9) is the guiding center radius, while the second term adds the length of the Larmor radius.

2.3.2 Maximal Reachable Radius as a Function of Electron Velocity

In order to find out how R_{max} is related to the electron kinetic energy (which can be expressed using the total speed V), eq. (2.9) has to be reformulated using the velocity V as the parameter rather than its components. The initial azimuthal speed is a part of the total speed, their ratio is

$$\sigma := \frac{V_\theta}{V} \in (-1, 1) \,, \tag{2.10}$$

α_{t} is the angle of the transition, fig. 2.1. For convenience, let

$$a_{\mathrm{t}} := \tan \alpha_{\mathrm{t}} = \frac{V_r}{V_z} \in (0, \infty) \,, \tag{2.11}$$

then, every component of the initial velocity can be expressed using V.

$$V_z = \sqrt{\frac{1 - \sigma^2}{1 + a_{\mathrm{t}}^2}} V \,, \tag{2.12a}$$

$$V_r = a_{\mathrm{t}} \sqrt{\frac{1 - \sigma^2}{1 + a_{\mathrm{t}}^2}} V \,, \tag{2.12b}$$

$$|\boldsymbol{V}_\perp| = \sqrt{V_\theta^2 + V_r^2} = \sqrt{\frac{\sigma^2 + a_{\mathrm{t}}^2}{1 + a_{\mathrm{t}}^2}} V \,. \tag{2.12c}$$

An example for the relative values of V_θ, V_r and V_z from a realistic MDC design is shown in fig. 2.3. (The design will be presented in section 2.4.)

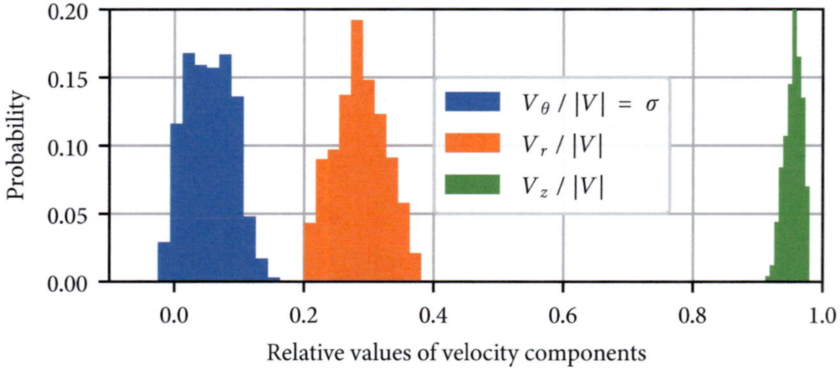

Fig. 2.3: Relative values of the initial velocity components in a realistic axisymmetric MDC design, taken at $z = 1.9$ m of the design in fig. 2.7

From fig. 2.3 one can see that a realistic average value of σ is approximately 0.09, which has a spread of ± 0.1. The value of σ is such low that it will be ideal for the demonstration of an axisymmetric MDC.

To shorten the final formulation, let

$$u := \frac{V}{\omega_c} > 0\,, \tag{2.13}$$

which is related to the electron kinetic energy and takes the relativistic mass (in ω_c) into account. Equation (2.9) can be formulated as a function of u:

$$R_{\text{max}} = \underbrace{\sqrt{\frac{\sigma^2 + a_t^2}{1 + a_t^2} u^2 + 2R_0 \sigma\, u + {R_0}^2}}_{r_{\text{gc}}} + \underbrace{\sqrt{\frac{\sigma^2 + a_t^2}{1 + a_t^2}}\, u}_{r_{\text{L}}}\ . \tag{2.14}$$

To investigate under which condition a faster electron will have a higher maximal radius in the region of the homogeneous B field, the two terms (r_{gc} and r_{L}) of eq. (2.14) will be checked separately.

2.3.3 Sorting of Guiding Centers

The r_{gc} term of eq. (2.14) has a quadratic polynomial under a square root:

$$r_{gc} = \sqrt{\frac{\sigma^2 + a_t^2}{1 + a_t^2} u^2 + 2R_0\sigma\, u + R_0{}^2}\ .$$

This parabola under the square root has a negative discriminant

$$\Delta = 4R_0{}^2\, \frac{a_t^2\,(\sigma^2 - 1)}{1 + a_t^2} < 0\ , \tag{2.15}$$

which means that the polynomial under the square root is always positive, as graphically shown in fig. 2.4 (which implies that r_{gc} cannot be zero or imaginary). The parabola has a turning point (minimum) at

$$u_{turn} = -R_0\sigma\, \frac{1 + a_t^2}{\sigma^2 + a_t^2}\ . \tag{2.16}$$

If the minimum of u (minimal electron kinetic energy) is on the right-hand side of this turning point, the guiding center increases monotonic with u, i.e. the guiding centers are modulated. The condition $u > u_{turn}$ can be written as

$$\frac{m_0}{q}\frac{1}{B_z} \min\{V\gamma\} > R_0|\sigma|\frac{1 + a_t^2}{\sigma^2 + a_t^2}\ . \tag{2.17}$$

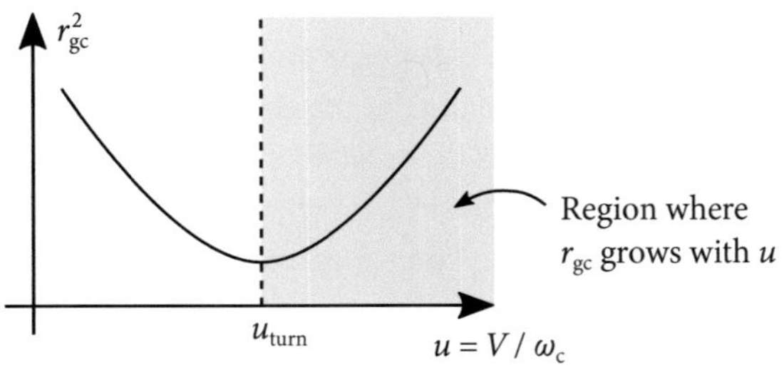

Fig. 2.4: Visualization of the r_{gc} term in eq. (2.14)

Here, $|\sigma|$ indicates the worst case. Hence, one finds the interval of B_z for the homogeneous magnetic field, which can cause a modulation of electron kinetic energy on r_{gc}:

$$|B_z| < \frac{m_0}{|q|} \min\{V\gamma\} \frac{1}{R_0} \frac{|\sigma| + \frac{a_t^2}{|\sigma|}}{1 + a_t^2} . \tag{2.18}$$

Since

$$|\sigma| + \frac{a_t^2}{|\sigma|} \geq 2a_t \quad \text{and} \quad |\sigma| < 1 ,$$

the conservative choice of the magnetic field should be in the range

$$|B_z| < \frac{m_0}{|q|} \frac{1}{R_0} \min\{V\gamma\} \cdot \begin{cases} \frac{2a_t}{1 + a_t^2} & a_t < 1 \\ 1 & a_t \geq 1 \end{cases} . \tag{2.19}$$

However, the magnetic field according to eq. (2.19) should be weak ($B_z \approx 1$ mT for $R_0 = 300$ mm and 8 keV minimum initial energy) and the non-adiabatic transition converts the small-orbit beam to a misaligned large-orbit one. Therefore, the sorting of guiding centers is not practical.

2.3.4 Sorting of Larmor Radii

Although the guiding centers may not be sorted nor separated, the second term of eq. (2.14) — the Larmor radius r_L is proportional to u for a given σ. Therefore, if the Larmor radii can be modulated by the electron kinetic energy, then, the maximal reachable radii are also modulated. However, there is a spread of σ, so that two electrons with the same kinetic energy (i.e. the same total speed V) may have different Larmor radii. The effects of the σ-spread will be discussed later in this section.

As shown in fig. 2.3, σ is usually low at the considered non-adiabatic transition, but a_t can be especially large. In the ideal case of a rapidly diverging magnetic flux (α_t approaching 90°), a_t is such large that the coefficient converges to unity

$$\lim_{a_t \to \infty} \sqrt{\frac{\sigma^2 + a_t^2}{1 + a_t^2}} u = u \,. \tag{2.20}$$

In this extreme case, the Larmor radius only depends on the kinetic energy. This effect is good, but not realistic.

For a realistic α_t (for example, $\alpha_t = 45°$ thus $a_t = \tan \alpha_t = 1$), the upper and lower boundaries of the Larmor radius can be calculated by assuming the minimal and maximal σ. Both boundaries increase with the electron velocity monotonically. Figure 2.5 shows the relation between the electron kinetic energy and the Larmor radius. Two instances of the magnetic field transitions:

- $\alpha_t = 45°$ (the reference value)
- $\alpha_t = 63.8°$ (the α_t in fig. 2.1)

are presented. The sorting of electrons is conditionally possible. For example, the lower boundary of the Larmor radius for a 60 keV electron is higher than the upper boundary of a 40 keV one. That means a 60 keV electron cannot have the same Larmor radius as a 40 keV one, independent of their azimuthal speeds at the end of the non-adiabatic transition. These electrons can be sorted. On the other hand, electrons with 40 and 50 keV kinetic energy do not have distinguishable maximal Larmor radii when $\alpha_t = 45°$. They cannot be well sorted by Larmor radius. But if the transition angle was increased to 63.8°, they would have different Larmor radii and are possible to be sorted.

Hence, it can be concluded from eq. (2.20) that a steeper transition angle (larger a_t) results in a sharper (better) modulation, as visualized by the thickness of both "bands" in fig. 2.5.

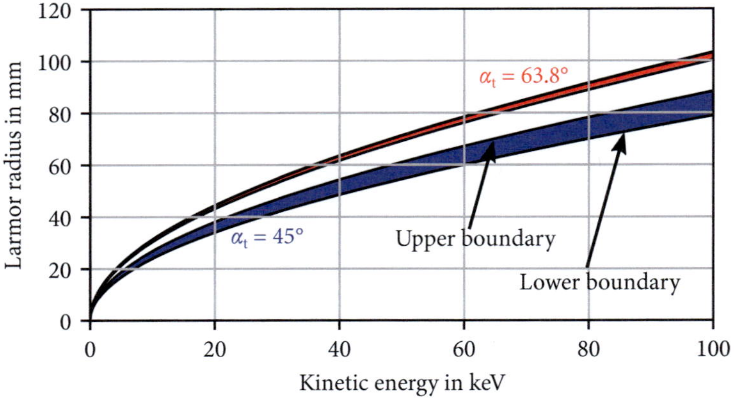

Fig. 2.5: Larmor radii after a non-adiabatic transition with the magnetic field in eq. (2.1) and electrons with $|\sigma| < 0.5$. Although there is a spread of the parameters (thickness of the bands), both upper and lower boundaries increase monotonically.

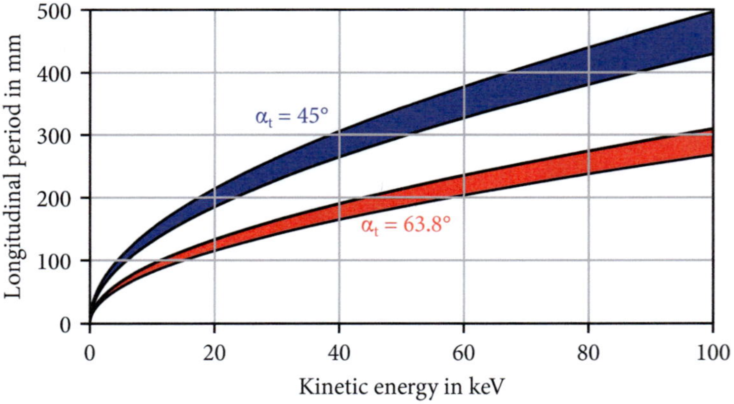

Fig. 2.6: Cyclotron longitudinal periods after a non-adiabatic transition with the magnetic field in eq. (2.1) and electrons assuming $|\sigma| < 0.5$

2.3.5 Sorting of Longitudinal Cyclotron Periods

The other modulated variable is the longitudinal period of the gyration, which is

$$\lambda = \frac{2\pi}{\omega_c} V_z \tag{2.21}$$

Substituting V_z with eq. (2.12a) results in

$$\lambda = 2\pi \sqrt{\frac{1-\sigma^2}{1+a_t^2}} u \,. \tag{2.22}$$

The relation between the longitudinal period and the electron kinetic energy is depicted in fig. 2.6. Opposite to the modulation of the Larmor radii, here, a smaller transition angle creates a larger longitudinal range in the same interval (abscissa) of electron kinetic energy, however, a larger angle results in a sharper energy sorting.

2.3.6 Properties of the Non-Adiabatic Transition Learned from this Analysis

In conclusion, the model abstracted from this kind of non-adiabatic magnetic transitions can provide the modulation (sorting) of the maximal reachable radius and longitudinal period length according to the electron kinetic energy. Following rules can be derived from the analysis

1. In order to clearly sort electron guiding centers within the homogeneous field, the magnetic field has to be low (typically 1 mT), such that it is not practical for a DEMO gyrotron.
2. In a more reasonable magnetic field of $B \approx 10$ mT, the spread of guiding centers can be neglected compared to the Larmor radii after the non-adiabatic transition. Therefore, the modulation of the maximal radius is equivalent to the modulation of Larmor radius.
3. For a particular α_t, the upper and lower boundaries of the post-transition Larmor radius increase monotonically with the electron kinetic energy. A steeper α_t causes a stronger modulation.

4. The upper and lower boundaries of longitudinal period length λ are also modulated with electron kinetic energy. To reduce the thermal load on electrodes, a wide region of longitudinal impaction is preferred, which however corresponds to a small a_t (this will be the case for the design later in section 2.4).

5. The spread of the azimuthal velocities causes an uncertainty in the modulation. Due to this reason, the resolution of the electron sorting cannot be so high as one would wish. This means that some electrons, which should be collected at the optimal stage, will be collected at the previous stages as well. Therefore, the ideal efficiency (fig. 1.13), which assumes a perfect collection, is hard to achieve. The resolution can be improved by increasing α_t.

2.4 Conceptual Collector Design and Basic Examinations

2.4.1 An Optimized Axisymmetric Collector Design for a 170 GHz 1 MW Gyrotron

The conceptual design and simulation of an axisymmetric MDC based on this principle for a 140 GHz gyrotrons was presented in [73]. To check the applicability of this principle for gyrotrons with a higher frequency, and to have the same reference gyrotron as well as the same reference SDC with the other design proposals presented in this work, an axisymmetric MDC proposal for the 170 GHz 1 MW gyrotron will be optimized and studied. The scaling of this concept to a DEMO gyrotron (170 GHz / 204 GHz 2 MW) will be discussed.

The magnetic field after the non-adiabatic transition is the most important parameter for the control of the Larmor radius. Electrons in the modulation region should have a sufficient dynamic of the Larmor radius (the difference between the largest r_L and the smallest r_L), in order to clearly distinguish their kinetic energies. Otherwise, if the radii of two electrons are close, the size and shape of an electrode should be very precise in order just to pick one packet of the electrons out of the sorted electron beam. According to eq. (1.18), if the magnetic field is too strong, the (absolute) dynamic range of trajectory oscillations is limited. On the other hand, if the magnetic field is chosen too weak, the collector will have a huge geometry or will require strong electromagnetic coils. From the examples in section 2.3, 10 mT can be a proper magnitude for the magnetic field. The dynamic range of the Larmor radii in section 2.3 can be 40 mm for an energy range of 60 keV. The magnetic field of the 170 GHz 1 MW gyrotron should be decreased to approximately 10 mT. A 204 GHz DEMO gyrotron will have a higher cavity magnetic field, therefore, a larger collector size or more complicated coils might be required for a 204 GHz gyrotron.

Additional solenoid coils (the green ones in fig. 2.7a) are placed around the collector to defocus the field. They pull those magnetic surfaces, which are outside of the electron beam, outwards. To further demagnetize the gyrotron static magnetic field without dramatically expanding the collector size, solenoid coils are inserted coaxially close to the collector center axis. These coils pull the magnetic flux surfaces, which are enclosed within the electron beam, inwards. As a result of the pulling and pushing from opposite sides, the flux density at the annular electron beam can be quickly reduced to approximately 10 mT. Such a transition of magnetic field might take place within a relatively short distance and the maximal radius of the electron beam (thus, the radius of the collector) could be acceptable.

The outside coils can be normal-conducting and are driven by Direct Current (DC). They replace the normal-conducting sweeping coils. As will be shown later, the sweeping of the electron beam is not necessary for the 1 MW gyrotron (the 2 MW case will be discussed later), because the non-adiabatic transition already spreads the electron beam so widely that the maximal static power load density is already below the time-averaged limitation (500 W/cm^2). The inner coils can be enhanced or replaced by ferromagnetic pole shoes [10, 11, 23]. In [23] it was shown that the system can be realized.

However, the solution of the magnetic field is not straightforward. The parameters of collector coils are not inversely solvable from an arbitrarily given flux profile. Therefore, the size, position and current of additional coils are optimized empirically. This is a drawback of the axisymmetric MDC design.

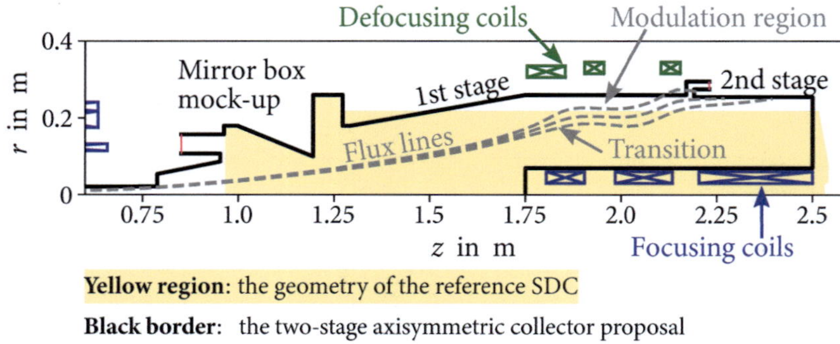

(a) Geometry, coils and magnetic field lines (magnetic flux surfaces)

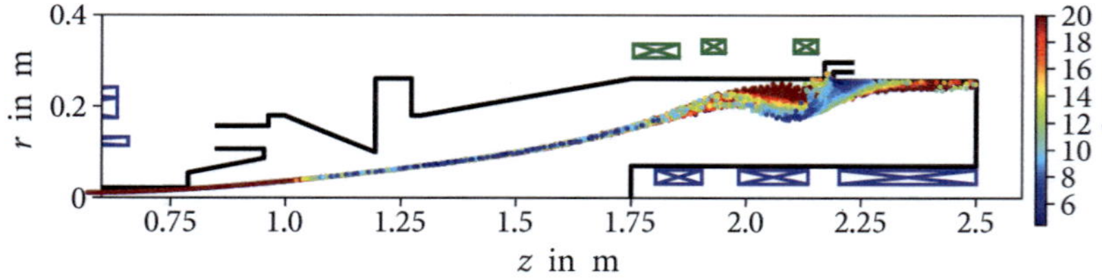

(b) Trajectory of the electron beam

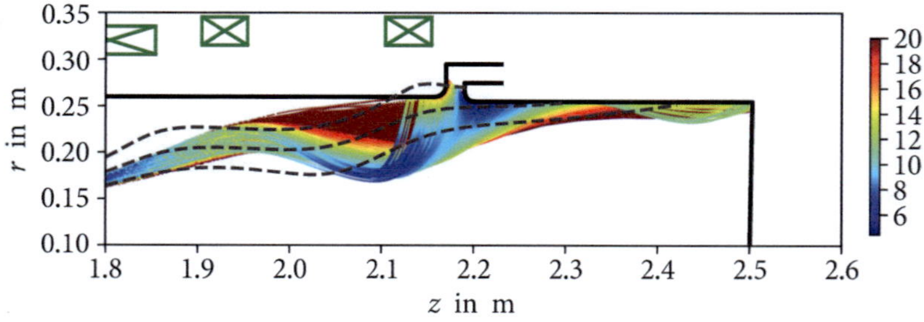

(c) A zoomed view of the trajectories in the region of the energy sorting

Fig. 2.7: Axisymmetric MDC design proposal for the 170 GHz 1 MW gyrotron

The optimized MDC design proposal is 10 cm shorter than reference SDC. The MDC design has an inner radius of 260 mm, which is larger than the 225 mm inner radius of the reference SDC. The reasons for the larger radius are

- to insert the center coils,
- to have enough space for the non-adiabatic transition,
- to enclose the larger Larmor radii after the transition.

The control of the trajectory is similar to the scheme shown in fig. 2.1. The magnetic field should modulate and sort electrons in absence of the electric field, whereas the electric field further decelerates the trajectories, which will impact the second stage. The difference to fig. 2.1 is that, due to the existence of the electric field, the angle of magnetic field is better to be adjusted to produce a higher radial component after the slow electrons gyrate approximately a half period (at $z \approx 2.05$ m). This part of the magnetic field steers the electron beam to electrodes at the second half period of the trajectories, in order to keep the shape of electrodes to be cylindrical.

The sorted electron beam impacts a large area of the collector wall. The gap between two stages locates exactly at the longitudinal position, where the sorted electron beam should be separated (see the depression voltages in table 1.2). At a first approximation, influences of secondary electrons are ignored. The simulation with depression voltages of 30 kV and 46 kV predicts a collector efficiency of approximately 73 % for this particular spent beam of the 170 GHz 1 MW gyrotron. (The simulated collector efficiency for the 140 GHz gyrotron is similar [73] to this one, while further studies are needed for the 204 GHz DEMO gyrotron.) The collector efficiency of 73 % can almost fulfill the goal of $\eta_{\text{col}} > 74\,\%$ for a $\eta_{\text{total}} > 60\,\%$.

2.4.2 Examination of the Effective Potential

The definition of the so-called "effective potential" can be found in the textbooks of plasma physics e.g. chapter 2.6 of [76]. It was mentioned in [9] that the effective potential would be a useful tool to assist the design of axisymmetric collectors, which include [7–12, 23, 73, 77, 78]. A brief description of its application is given in the following paragraphs.

The effective potential can be derived from the total energy (Hamiltonian $\mathcal{H}$) of a charged particle in a non-relativistic axisymmetric system:

$$\mathcal{H} = \frac{p_z^2}{2\,m} + \frac{p_r^2}{2\,m} + \frac{1}{2\,m}\left(\frac{p_\theta - q\,r\,A_\theta}{r}\right)^2 + q\,U\,, \tag{2.23}$$

where the effective potential Φ is the sum of the last two terms in eq. (2.23): the energy in the azimuthal motion (therefore, only for a non-relativistic system) plus the electric potential energy of the particle:

$$\Phi = \frac{1}{2\,m}\left(\frac{p_\theta - q\,r\,A_\theta}{r}\right)^2 + q\,U\,, \tag{2.24}$$

$p_{\{z,r,\theta\}}$ are the components of the canonical momentum. Busch's theorem in eq. (1.24) points out, that the canonical angular momentum p_θ of each individual electron is constant, as far as the electric potential and magnetic vector potential axisymmetric. In this way the term for the "azimuthal energy" can be calculated for the known r and $A_\theta(z,r)$ at any position in the system. An electron can be maximally decelerated until both z and r components of its velocity vanish, i.e. at the moment when $\mathcal{H} = \Phi$; then, the electron cannot move forward and would turn back. This is the boundary, where an electron with a total energy of $e\,\Phi$ may reach in the z-r projection.

Figure 2.8 shows the effective potential of this MDC design. The value of effective potential is individual for each electron, but for the application of such an MDC, the effective potentials of electrons are similar. Therefore, a random electron is chosen from the beam for the evaluation of Φ. Equipotential lines with 40, 50 and 60 kV are marked in the diagram. For example, only electrons with higher than 40 keV initial energy can move beyond the 40 kV equipotential line.

Comparing fig. 2.8 and fig. 2.7b side by side, one recognizes, that the potential model is not deterministic. It only indicates the theoretical farthest reachable location of electrons with that initial energy; however, the electrons do not reach every point on potential boundary; i.e. it is a sufficient but not a necessary condition for the electron trajectory control. Therefore, the effective potential is only an auxiliary parameter for the design of an axisymmetric MDC.

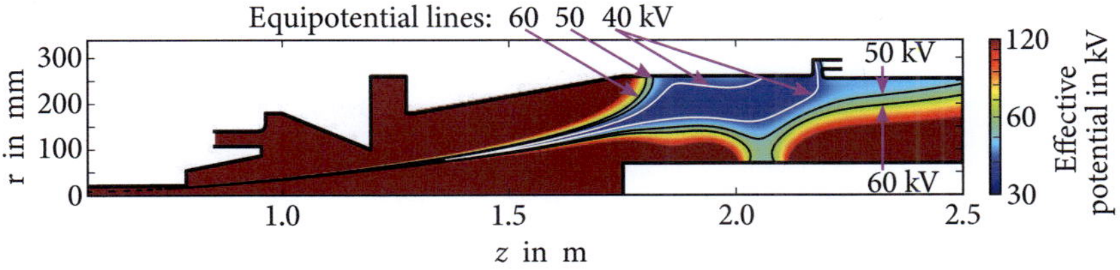

Fig. 2.8: Effective potential

2.4.3 Examination of the Energy Sorting Quality Using Mono-Energetic Electron Beams

A set of simulations using mono-energetic, low-perveance electron beams can examine how well and how sharp an MDC theoretically sort the beam electrons. To simplify the beam sampling, the injected macro electrons do not have initial transverse velocities. The electrons will gain transverse velocities during the demagnetization process.

Since the conceptual design in fig. 2.7 was optimized considering the space charges, to reproduce the same situation as with a realistic electron beam (having the correct current and a realistic energy spectrum), the electric field in each examination is imported from the simulation of the 45 A electron beam with a realistic energy distribution. The final electric field consists of the statically applied electric field and the contribution from the space charges of the 45 A realistic beam, whereas the space charges of an injected mono-energetic electron beam are not considered.

Figure 2.9 shows the relative current on each stage as a function of the electron initial kinetic energy. Since the examination includes the space charges from the realistic beam, the mono-energetic electrons also have the same initial potential energy as in that assumed one. Therefore, the second stage at a depression voltage 46 kV in this example can even collect those electrons, which only have 40 keV initial kinetic energy. From fig. 2.9 one can see that electrons with less than 39 keV energy are collected only at the first stage, while the ones between 44 and 70 keV are collected only at the second stage. The sorting is ambiguous in the intermediate interval from 39 to 44 keV. Those electrons are spread to both stages.

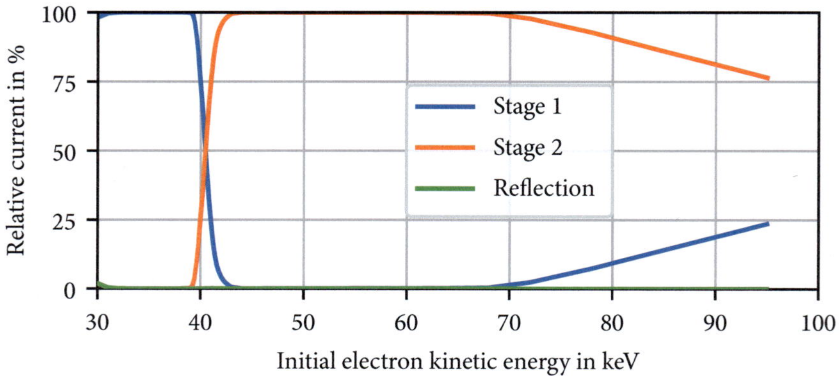

Fig. 2.9: Examination of the sorting quality using mono-energetic electron beams for the axisymmetric MDC design

The intermediate interval should be as narrow as possible, as a sharp sorting can cause a high collector efficiency. However, the energy sorting cannot be done precisely according to the theory in section 2.3. Hence, it is not trivial to shrink the intermediate region in this kind of MDC.

Electrons with an initial energy above 70 keV are partly collected at the first stage. The reason can be explained from fig. 2.1, that for an electron with a much higher energy, its periodic length can be so long that it may impact the first stage already within the first half period of its cyclotron motion. Fortunately, the number of electrons with very high energy is not large in a realistic spent electron beam. Collecting them at the first stage does not decrease the collector efficiency too much. Even if the cavity mode collapses or could not start up, the electrons have a high kinetic energy and would be in the worst case totally collected at the first electrode with a non-optimal depression voltage like an SDC.

2.5 Assessment of the Critical Parameters

In this section, the parameters of this axisymmetric MDC design will be checked according to the expectations and constraints given in section 1.5.

2.5.1 Achievable Collector Size and Mechanical Complexity

This axisymmetric MDC design for the 170 GHz 1 MW gyrotron is 1.23 m long, which is similar to the reference SDC with beam sweeping. It has an inner radius of approximately 250 mm. Considering the isolation between two stages, the outer radius of the MDC is expected to be larger than 300 mm. Due to the strong non-adiabatic transition of the magnetic field, this MDC design should be much compacter than an up-scaled version of the axisymmetric MDC variant in fig. 1.18, where the electrostatic field dominates the energy sorting. (That design for the 110 GHz gyrotron has already a larger radius than this one.)

The electrodes in this design are regular cylinders. However, in order to weaken the magnetic field, there are in total six additional coils, which are not only outside the collector but also inserted coaxially. Like [10, 11], the inner coils might be replaced or enhanced by pole shoes.

2.5.2 Expected Collector Efficiency and Unwanted Back-Stream Current

If the influence of secondary electrons is ignored, the simulation of the two-stage depressed collector shows a collector efficiency of $\eta_{\mathrm{col}} = 73\,\%$, which almost reaches the goal of 74 % in order to have an $\eta_{\mathrm{total}} > 60\,\%$. There is no back-streamed primary electron observed in the simulation. (However, if secondary electrons are considered in the simulation, 22.5 mA are back streamed to the cavity.) As there is no separation of electron orbits, this concept cannot be extended to have more than two stages for a higher efficiency, at least not for this 170 GHz 1 MW reference gyrotron within the assumed size limitation.

In the operation environment of a fusion gyrotron, the stray magnetic fields from the tokamak and from the neighboring gyrotrons overlap with the gyrotron magnetic field itself. This perturbation field may affect the operation of the gyrotron collector [79, 80]. The longitudinal component of the magnetic field can be compensated by the collector solenoid coils, while the transverse field component can tilt the electron beam. If the electron beam is tilted away from the assumed position, the collector would not operate as it was expected from the design. The influence of the transverse field component will be presented the next paragraph.

Figure 2.10 shows the efficiency η_{col} and the back-streamed current I_{ref} versus the magnitude of a transverse perturbation magnetic field $B_{\perp}$. In such an axisymmetric system, any direction of the global perturbation field produces the same effect. The simulation takes the space charge effect into account, whereas secondary electrons are ignored. According to [79, 80], a flux density up to 0.5 mT should be considered for this study. (For a comparison, the Earth magnetic field at the Earth surface is circa 25–65 µT [81].) A field of 0.5 mT causes 5.4 mA back-stream current, that is less than the order of the back-stream current caused by secondary electrons. However, the collector efficiency is reduced to 65 %. For a higher field strength, the MDC performance falls to the level of an SDC. Hence, the sensitivity of external perturbation magnetic field can be a weakness of such axisymmetric MDC systems.

Due to the misalignment in the experiments, the electron beam might have up to 0.5 mm displacement in the cavity [82]. After the magnetic decompression, the displacement in the collector is of several orders larger than in the cavity. Here, the collector efficiency and reflected current are examined by simulations with a variable beam displacement. Results are shown in fig. 2.11. In these simulations, only the beam position is transversely shifted by 0.1–0.9 mm distances, while the magnetic field and the collector structure are assumed to be correctly aligned. It is found that a displacement of 0.5 mm in reduces the collector efficiency by 3 %, which is minor compared to the other effects. Additionally, the misalignment introduces up to 8 mA of back-streamed current, which is also less than the one caused by secondary electrons.

As mentioned before, that there is a magnetic flux density of approximately 10 mT in this MDC design. The electric force from the charges in the electron beam can influence the trajectories in such a flux density. Therefore, the design and optimization of an axisymmetric MDC should take into account the space charges of the electron beam. The space charge effect is related to the current of the electron beam. To investigate the sensitivity of the beam current, simulations with a variable beam current are performed.

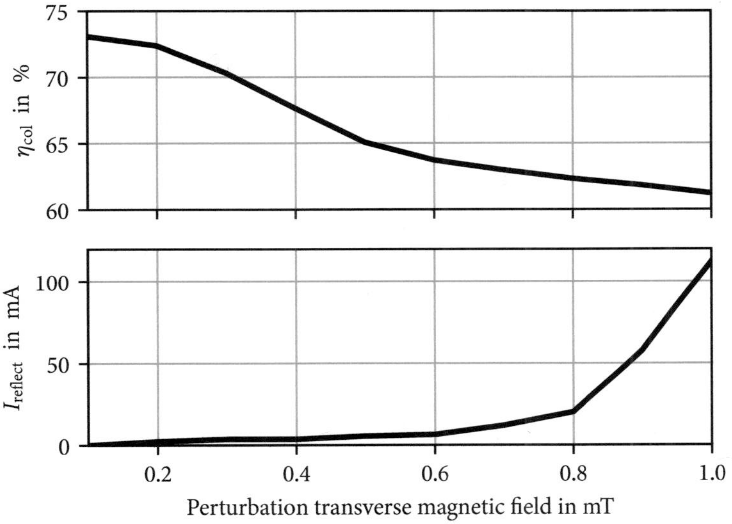

Fig. 2.10: Sensitivity of the perturbation radial magnetic field

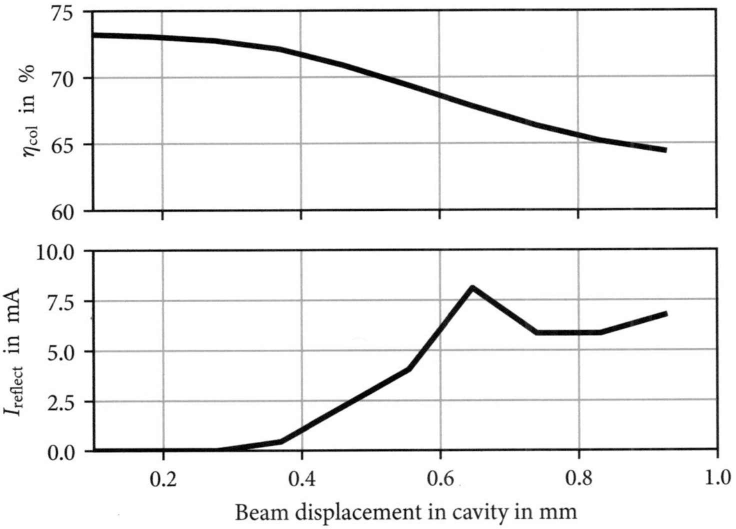

Fig. 2.11: Sensitivity of the beam displacement in the cavity

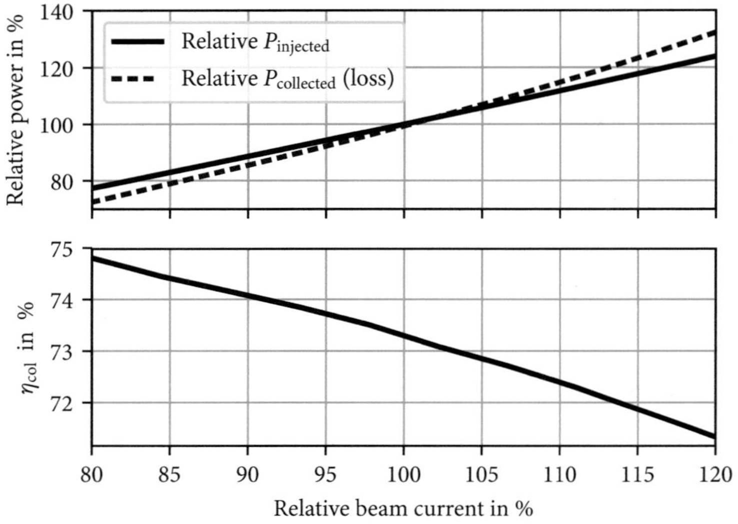

Fig. 2.12: Sensitivity of electron beam current

Figure 2.12 shows the simulated collector efficiency as a function of the beam current. The power and current in fig. 2.12 are normalized to the 45 A beam. Although the MDC performance depends on the space charge, at least for a small current range of ±20 %, the collected power at the stages increases almost linearly with the beam current. Using the same depression voltages in all simulations, η_{col} varies ±2 % absolutely. No primary electrons are reflected due to the variation of the beam current.

The effect of secondary electrons is the last point to be checked in this section. The reduction of η_{col} by secondary electrons are unavoidable especially in this type of MDC concept, for following reasons:

1. There is no real separation of electron trajectories according to Busch's theorem. The electron beam is spread by the magnetic field continuously over a certain area at the collector wall.
2. As the beam spread is spatially continuous, the two electrodes are close to each other, thus, the secondary electrons can easily form a (negative) current flowing from the second stage to the first one.

3. For an MDC consisting of two stages, the optimal depression voltage of the second stage should be close to the peak in the energy spectrum, as shown in fig. 1.12. Since the electron beam is approximately energy-sorted on the collector wall, the gap for the separation of the stages (shown in fig. 2.7) is just at the position in front of the peak of the beam current distribution. Hence, the beam current, which impacts the front part of the second stage is particularly high (this can also be seen later from the estimation of the power load). A higher impacting current of the primary beam causes a higher current of emitted secondary electrons. They are the secondary electrons, which form the main part of the inter-stage current.
4. The magnetic field has a flat angle to the electrodes, in order to spread the primary beam over a large area. The direction of the magnetic field allows the "path" of the inter-stage current to be relatively straight and short. This also increases the chance of forming the unwanted secondary electron current.

Figure 2.13a shows the inter-stage current, which consists only of secondary electrons. This current reduces the collector efficiency by 7.7 % and will finally render the gyrotron overall efficiency (η_{total}) to be only 55 %. Such an efficiency reduction of 6 % to 10 % is typical for an axisymmetric gyrotron MDC. For instance, the collector efficiency in [10] is reduced by from 68 % to 60 %, while in [65] from 90.4 % to 82.9 %.

Various methods to suppress the secondary electrons are introduced in section 1.3.2. The method to groove the electrode surface macroscopically, which does not require any unknown surface parameters of other materials, will be checked. (A similar study was also reported in [78].) Based on the original design, grooving the front part of the second stage can gain 2.5 points of the efficiency loss at minimum. The grooves are 5 mm wide and 5 mm deep. The distance between two grooves is also 5 mm. (These parameters are arbitrary and may not be the optimum.) Theoretically, a larger ratio between the depth and the width has a better suppression effect [45]. Figure 2.13b shows graphically that the inter-stage current is reduced when the surface is grooved. As given in table 2.1, the inter-stage current in this case is reduced by 3.7 A.

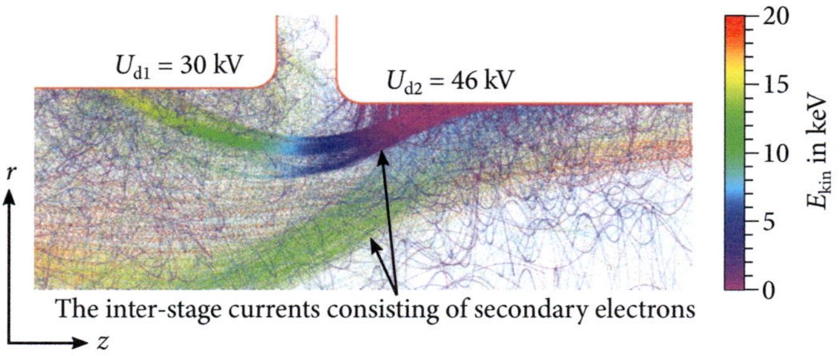

(a) The original design

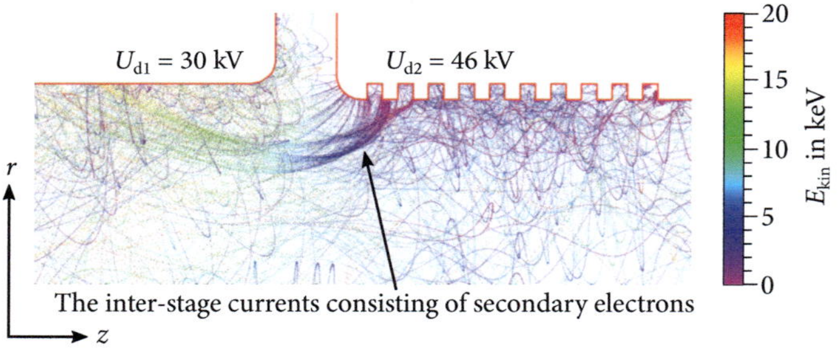

(b) Grooved second stage

Fig. 2.13: Trajectories of secondary electrons

Table 2.1: Influence of secondary electrons on efficiency and currents

	Straight* electrode	Grooved* electrode	Without secondary electrons
η_{col}	65.5 %	68.0 %	73.3 %
I_{ref} (mA)	22.5	13.5	0
I_{b1} (A)	17.78	14.03	6.25
I_{b2} (A)	27.20	30.96	38.75

* The simulation of the secondary electron emission involves random numbers for the emitted electron energy and angle. These are the representative values when the same random seed is applied.

The simulations considering secondary electrons only terminate when the last electron in the simulation region is collected, in order to obtain a balance of the beam current and to make sure that there is no trapped electron. To terminate the simulation within a reasonable simulation time, maximal 10 generations of secondary electrons are emitted in the simulation. The back-stream current only consists of secondary electrons. The grooves can also reduce the back-stream current, see table 2.1 for a comparison.

2.5.3 Maximal Expected Power Load Density on the Critical Parts

Figure 2.14 shows the results from the ESRAY [38] simulation for the power load density on the collector wall. The thick border is the axisymmetric profile of the collector inner surface. The line segments perpendicular to the border stand for the power density of the electron bombardment. A longer segment means that the power density is higher at that location. In brief, the power load density is in the acceptable range, i.e. everywhere on the electrodes are below 500 W/ cm^2. Trajectory sweeping is not necessary.

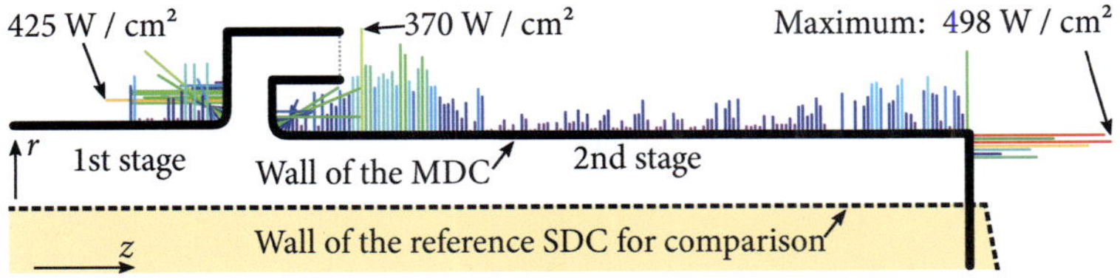

Fig. 2.14: Power load density on the wall of the axisymmetric MDC proposal for the 170 GHz 1 MW gyrotron

A non-adiabatic transition does not always sort electrons in the expected way, but it can spread the electron beam spatially, in particular when the final magnetic field allows a large Larmor radius like in this conceptual design. With a proper impinging angle, the beam electrons are spread over a large area on the collector wall. For example, the electron beam in this simulation is spread over a cylindrical area of 6500 cm^2 according to fig. 2.7c. Aside from the extremely large spread, the second stage has a more negative potential than the first one. The depression voltage between two stages further decelerates electrons and reduces the load at the second stage. Due to these reasons, the power load of this axisymmetric MDC design is within the currently allowed limit of 500 W/ cm^2, if the influence of secondary electrons is not considered. The three highest loadings are:

- The maximal power load density of 498 W/ cm^2 appears at the end. Since the geometry of this MDC model is truncated at a length similar to the reference SDC (100 mm shorter than the SDC), the high-energetic electrons impact the "cap" of the second stage perpendicularly, as fig. 2.7c shows.
- The second highest spot with 425 W/ cm^2 is located at the first stage inside the gap. Electrons entering the gap have higher energy than those collected at the lowest z due to the energy sorting. The electric field pushes almost all electrons, which enter the gap, toward the first stage. Therefore, a high power load density exists at the side of the first stage inside the gap.
- The third highest power load density appears at the front part of the second stage, since the peak (i.e. high current) of the energy spectrum is at approximately 50 keV, as shown in fig. 1.12.

The energy carried by the secondary electrons will be redistributed, when those electrons are re-collected. Therefore, the power load density differs from the predicted value by the ideal simulation without secondary electrons. The maximal power load densities, which also take into account the power redistribution by the secondary electrons, are given in table 2.2. The inter-stage current formed by the secondary electrons increases the maximal power load density at the first stage. The one at the second stage is also increased due to the bouncing of secondary electrons.

Grooving the second electrode changes the maximal load densities at both stages. At the second stage, the load is further increased, since some electrons impact the edges of grooves with almost perpendicular angles. The maximal load density at the first stage is lower than with a straight second stage, because the inter-stage current is lower. In addition, the secondary electrons redistribute the power.

Table 2.2: Influence of secondary electrons on the power load density

	Straight electrode	Grooved electrode	Without secondary electrons
Maximal load at stage 1 (W/cm^2)	521	342	425
Maximal load at stage 2 (W/cm^2)	660	712*	498

* can be further optimized by tuning the angle of fields and angle of impacting.

2.5.4 Summary

Axisymmetric MDC is the classical approach to start with. The existing designs [7–11, 23, 77, 78] permitting only a low magnetic field can be hardly up-scaled to the gyrotrons, in which the electron beam encloses a higher flux. The new sub-type of axisymmetric MDC presented in this chapter sorts electrons by their cyclotron phase, therefore, it is more suitable for the future fusion gyrotrons with higher frequencies. To demonstrate the properties of this concept, an axisymmetric MDC designed for the 170 GHz 1 MW gyrotron is presented.

The major advantages of the axisymmetric MDC are the simplicity of the symmetry and a large impacting area of the electron beam. Like the designs of the other axisymmetric MDCs, coils or pole shoes are inserted inside the beam, in order to control the collector size. This increases the complexity.

The main limitation of such an axisymmetric MDC is that there is no real separation of electron orbits, according to Busch's theorem. Without a real separation of the orbits, the energy-sorting is theoretically imperfect and thus it limits the achievable η_{col} for a two-stage collector. Extending to an MDC with more than two stages is challenging without a real orbit separation. Also, secondary electrons can easily move between stages, which causes a reduction of the collector efficiency.

For a 2 MW DEMO gyrotron with doubled current (up to 90 A) of that in the 170 GHz 1 MW gyrotron, it can be foreseen that such an MDC will have 1 kW/cm^2 power load density for the same maximal radius. As the beam sweeping is not possible and the concept is not able to be extended to three or more stages, the collector radius (thus the area of beam impaction) should be doubled in order to achieve the target of 500 W/cm^2, which requires a collector inner radius of approximately 0.6 m.

3 Proposals for Non-Axisymmetric Collector Concepts Based on $E \times B$ Drift

3.1 Advances to the State-of-the-Art in Brief

The first gyrotron MDC design proposal based on the $\boldsymbol{E}\times\boldsymbol{B}$ drift was invented in 2008 by Pagonakis et al. [68]. That conceptual investigation was limited due to the lack of a three-dimensional simulation tool. As a result, an infinite number of collector stages was considered in that proposal. Today, more convenient tools for the particle simulation, such as the CST Particle Studio, are available. With the modern simulation tools, those conceptual designs, which require a full three-dimensional modeling, are able to be investigated in addition to the conventional axisymmetric concept.

The first design proposal for the $\boldsymbol{E}\times\boldsymbol{B}$ type of gyrotron collector [68] was one of the numerous possibilities to use the $\boldsymbol{E}\times\boldsymbol{B}$ drift for the energy-sorting of electrons. Prior to the presented work, it was unknown, what kind of possibilities there would be. In the presented work, various MDC concepts based on the $\boldsymbol{E}\times\boldsymbol{B}$ drift are invented. They are numerically evaluated and categorized for the first time. Exploring a proper way to collect the gyrotron small-orbits annular electron beam using the $\boldsymbol{E}\times\boldsymbol{B}$ drift is non-trivial. Hence, there is a systematic study of the possibilities for the application of the $\boldsymbol{E}\times\boldsymbol{B}$ drift. In this chapter, the conceptual proposals for the implementation of an $\boldsymbol{E}\times\boldsymbol{B}$ MDC are presented.

Via the analysis and simulation of each proposal, one can have a deep understanding of the properties of the operation principle, in order to further optimize the designs. For example,

- The theory of the relation between the configuration of coils and the shape of magnetic flux surfaces (field lines) is developed, in order to find the simplest configuration of coils for the sheet-beam transformation.

- From the analysis one can estimate the minimal collector size limited by the electron sorting mechanism.
- The back-streaming beam current can be minimized according to the properties of the reflected electrons.

Based on the understanding of the operation principles and the simulated performance of the MDC design proposals, the most promising MDC proposal is further improved and its design for the mechanical engineering is presented.

3.2 Possibilities to Use the *E*×*B* Drift in a Gyrotron Collector

There are several possibilities to create an $\boldsymbol{E}\times\boldsymbol{B}$ drift. Generally, the drift velocity can be planar (Cartesian coordinates) or radial (cylindrical coordinates). Both types can be further subdivided as shown in fig. 3.1.

In this section the basic properties of each scheme are presented. The maximal drift distance D (see fig. 1.20) is estimated for the ideal cases, where the components of electric and magnetic field are assumed to be constant.

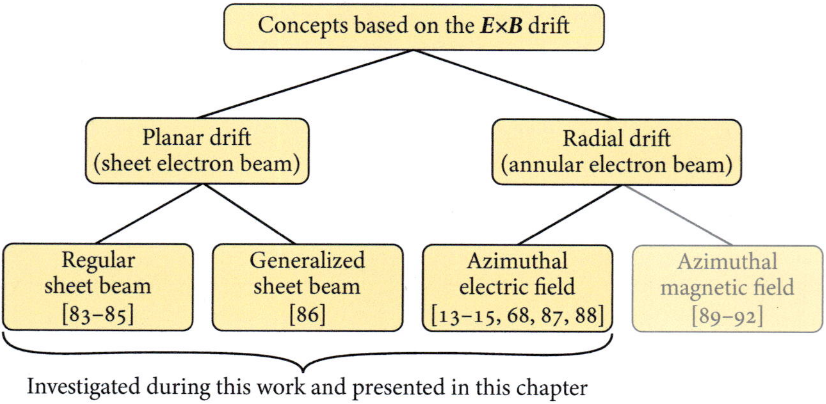

Fig. 3.1: Category of MDC schemes based on the $\boldsymbol{E}\times\boldsymbol{B}$ drift

3.2.1 Using a Planar Drift

A straightforward idea to collect a sheet electron beam by **E**×**B** drift is to extrude the electrodes in fig. 1.20 (on page 38) by an angle ϑ. The scheme is depicted in fig. 3.2. This idea introduces an $E_\perp$. Another possibility is to introduce a $B_\perp$, instead of $E_\perp$. Actually, both cases are identical, as the coordinates of one case can be rotated to become the other. Therefore, it is sufficient to analyze only the former case for the understanding of the planar drift properties.

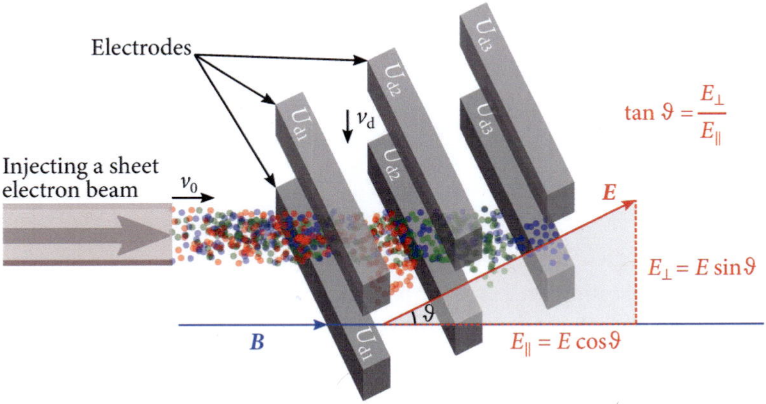

Fig. 3.2: Scheme of an MDC for a sheet electron beam

An electron needs time t to be longitudinally decelerated from $v_\parallel = v_0$ to $v_\parallel = 0$:

$$v_0 = \int_0^t \frac{q}{m} E_\parallel \, \mathrm{d}t \; . \tag{3.1}$$

Here, the sign of $qE_\parallel$ is unimportant, thus it can be safely assumed to be positive. In a non-relativistic system with a constant $E_\parallel$, that integral is

$$v_0 = \frac{q\,E_\parallel}{m} t \; . \tag{3.2}$$

The definition of the drift distance is given as follows. If there was no obstacle (e.g. an electrode) on the path of an electron to capture (collect) that electron, the electron would be decelerated until $v_{\parallel} = 0$ and then accelerated backwards by the electric field. At the moment when it returns back to the initial longitudinal position where it started, the electron has been exposed in the **E**×**B** drift for time duration t_d. This t_d includes the time of the deceleration and the backward acceleration (therefore, it has a coefficient of two):

$$t_d = 2\frac{m}{q}\frac{v_0}{E_{\parallel}} \,. \tag{3.3}$$

The drift distance D, as marked in fig. 1.20, is defined as the final transverse displacement of an electron. It is the product of the drift velocity $\boldsymbol{v}_{\mathrm{d}}$ from eq. (1.33) and the drift time t_d from eq. (3.3):

$$D = |\boldsymbol{v}_{\mathrm{d}}|\, t_d = \frac{2\, m\, v_0}{q\, B_{\parallel}}\frac{E_{\perp}}{E_{\parallel}} = \frac{2\, m\, v_0}{q\, B_{\parallel}} \tan\vartheta \,, \tag{3.4}$$

where $\vartheta < 90°$ (because a positive B_z is chosen in the gyrotron) is the angle between the electric field and the magnetic field, as graphically presented in fig. 3.2. An interesting property of the drift distance D in eq. (3.4) is that this distance is independent of the magnitude of the electric field, as far as the field is homogeneous like in this assumption. The independence of the electric field strength means that the distance between two collector stages can be theoretically arbitrary. The stages do not need to be placed such close to each other as in the axisymmetric concept in chapter 2. This is a significant advantage, since

1. The orbits of electrons can be separated over a large distance. If there would be any secondary electrons, which could not be properly re-collected by the drift (especially at the last stage), the electric field at that position can be tuned locally such that the inter-stage current of secondary electrons can be minimized.

2. As the electrodes are not necessary to be placed very close to each other as in chapter 2, a collector based on the **E**×**B** drift can have more than two stages to achieve a higher efficiency.

3. The independence of electrode distances can allow a certain tolerance in the design and fabrication.

In principle, a large D is preferred in the design to make the drift effective. D depends on three parameters, as shown in eq. (3.4), thus, there are three possibilities to achieve a large D:

1. to increase the initial electron velocity v_0,
2. to decrease the magnetic field $B_{\parallel}$,
3. to increase the angle ϑ.

They have the following side effects:

1. There are two options to increase v_0 without changing other parameters. The first option is to apply lower depression voltages, which would reduce the MDC efficiency. The second option is to adiabatically unwind more $v_{\perp}$ to $v_{\parallel}$, which requires a larger space. As the remained $v_{\perp}$ in an ***E***×***B*** MDC is already low, a further reduction of $v_{\perp}$ might not be practical.
2. The magnetic field can be reduced in the ways either to place the collector further away from the gyrotron main coils, or to use collector coils. Both can increase the collector size and complexity. Furthermore, a too-weak magnetic field does not have enough confinement of the electrons, such that the overall performance of the ***E***×***B*** collector may decrease. Therefore, a compromise has to be made on the choice of $B_{\parallel}$.
3. Increasing the angle ϑ between the electric field and the magnetic field means to place electrodes more parallel to the ***B***-field as shown in fig. 3.3. However, that will also increase the length of the collector.

Hence, the choice of D has to balance the effectiveness of the drift, the collector size and its complexity. A design proposal to obtain a high drift without significant increase of the collector size will be presented in section 3.7.2.

3.2.2 Using a Radial Drift

In a gyrotron depressed collector, there should be always a longitudinal (axial) component of magnetic field B_z from the gyrotron main SC magnet and a longitudinal component of the electric field E_z for the deceleration of the beam electrons. Generally, there are two approaches to create a radial drift: either to apply an azimuthal electric field E_θ, or to apply an azimuthal magnetic field B_θ. Both approaches are depicted in figs. 3.4a and 3.4b, respectively.

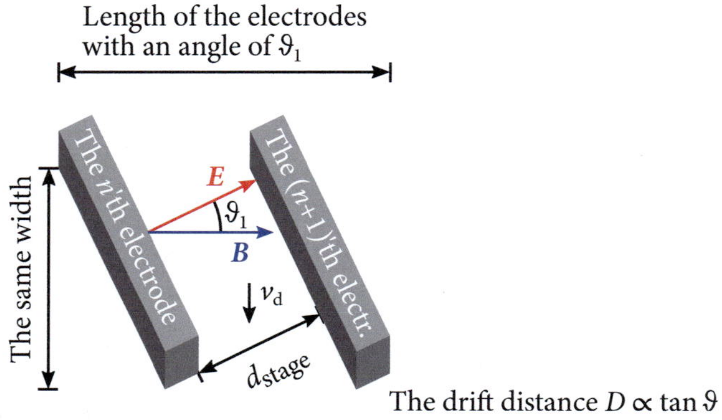

(a) Small angle $\vartheta = \vartheta_1$

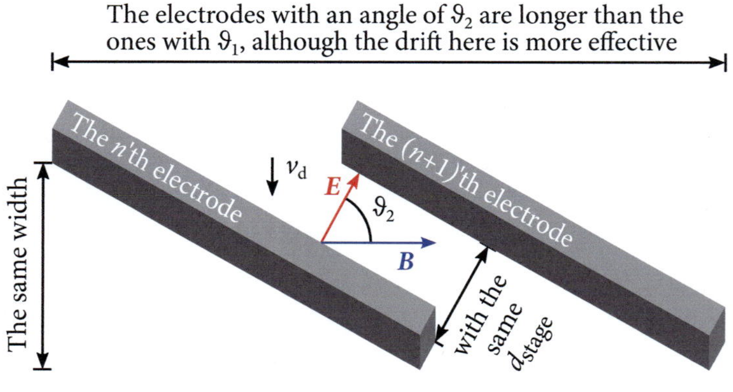

(b) Large angle $\vartheta = \vartheta_2$

Fig. 3.3: Increasing the angle ϑ between the electric field and the magnetic field will also increase the collector length

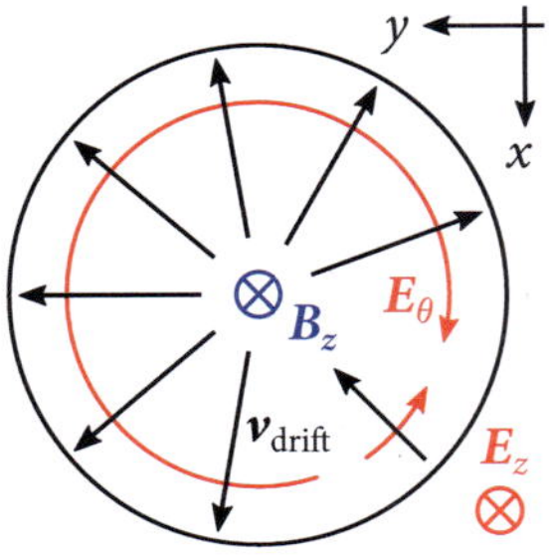

(a) Azimuthal electric field

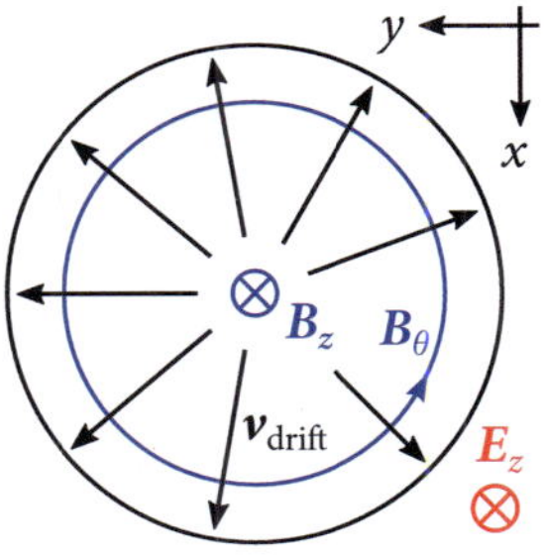

(b) Azimuthal magnetic field

Fig. 3.4: Cross sections of two possibilities to produce radial **E**×**B** drifts

Azimuthal Electric Field

The azimuthal electric field E_θ and the longitudinal magnetic field B_z result in an **E**×**B** drift of the beam electrons, as shown in fig. 3.4a.

As pointed out in section 1.4.3, that E_θ in a closed azimuthal path cannot always be in one direction due to Faraday's law. Wherever E_θ changes its sign, the drift velocity also turns to the opposite direction. The turning of the drift direction is unwanted and has to be specially handled. In fig. 3.4a, the direction of the drift is flipped at the right-bottom corner, while in other configurations, this angle can be as large as a half circle [68] or may repeat many times azimuthally (will be presented in section 3.7.2).

The drift distance D for this type of drift can be calculated using eq. (3.4), where in that equation $E_\perp = E_\theta$, $E_\parallel = E_z$ and $B_\parallel = B_z$.

Azimuthal Magnetic Field

Figure 3.4b shows the schematic for the case, where the **E**×**B** drift is created from the azimuthal component of the magnetic field B_θ and the longitudinal component of the electric field E_z.

In order to create a B_θ, there should be an axial current enclosed within the electron beam. For example, to create an azimuthal magnetic field of $B_\theta = 10\,\mathrm{mT}$ at the beam radius of $r_{\mathrm{gc}} = 200\,\mathrm{mm}$, an axial current of

$$I = \frac{2\pi\, r\, B_\theta}{\mu_0} = 10\,\mathrm{kA}\,(\cdot\,\mathrm{turns}) \tag{3.5}$$

is required. Equation (3.5) is the solution of the Biot-Savart integral of an axial long wire. If this current loops back from the outside of the collector (see the coil configuration in appendix B.1), a current in the order $5\,\mathrm{kA}\cdot$ turns is still required. This current is several orders higher than the electron beam current. Such an MDC was proposed by Louksha et al. [92].

The radial drift distance using the azimuthal magnetic field (under the assumption of a locally constant B_θ) is

$$D = |\boldsymbol{v}_\mathrm{d}|\, t_d = \frac{2\, m\, v_0}{q} \frac{B_\theta}{B_\theta^2 + B_z^2} = \frac{2\, m\, v_0}{q\, B} \frac{B_\theta}{B}\,. \tag{3.6}$$

Equation (3.6) is also independent of the electric field strength. However, D depends indirectly on the ratio between B_θ and B_z. Comparing eq. (3.4) and eq. (3.6) it is found that

$$\left|\frac{2\, m\, v_0}{q\, B} \frac{B_\theta}{B}\right| < \left|\frac{2\, m\, v_0}{q\, B}\right| < \left|\frac{2\, m\, v_0}{q\, B_z}\right| \tag{3.7}$$

It means that the drift produced by the B_θ can be less effective than the drift produced by an E_θ.

The best-case scenario of this type of drift can be calculated as follows. Since the B_z field in a gyrotron collector practically exists due to the gyrotron main SC magnet, this field component is inconvenient to vary for the whole collector length. Therefore, it is assumed here that a fixed B_z is given. On the other hand, the B_θ field is externally applied and is variable by changing the coil current of eq. (3.5). The drift distance in eq. (3.6) has a maximum when

$$\frac{\mathrm{d}D}{\mathrm{d}B_\theta} = \frac{1}{B_\theta^2 + B_z^2} - \frac{2\, B_\theta^2}{(B_\theta^2 + B_z^2)^2} = 0\,, \tag{3.8}$$

which has the solution $B_\theta = \pm B_z$. Choosing these values results in the maximum of the curve in fig. 3.5. Hence, in order to create an effective drift using a B_θ field, the choice of B_θ can be critical and is related to the value of B_z.

To summary, the MDC concept using B_θ has the following drawbacks:

1. It is challenging to insert a current inside the electron beam.
2. The drift may not be as effective as the drift produced by an E_θ.

Therefore, the investigation of an MDC based on this type of drift is excluded from this work.

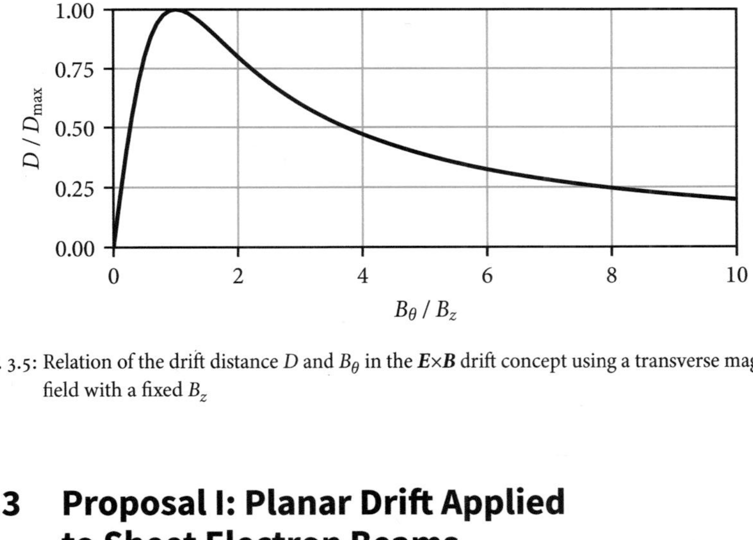

Fig. 3.5: Relation of the drift distance D and B_θ in the $\boldsymbol{E}\times\boldsymbol{B}$ drift concept using a transverse magnetic field with a fixed B_z

3.3 Proposal I: Planar Drift Applied to Sheet Electron Beams

3.3.1 Creation of Sheet Electron Beams

The planar drift is the first concept to be considered in this chapter, because the creation of a planar (even uniform) E-field and B-field required for this drift is in principle more feasible compared to the radial drift. Such an MDC requires a special "sheet beam injector", as shown in fig. 3.2, which is the component to transform the spent annular electron beam to one or more sheet electron beams.

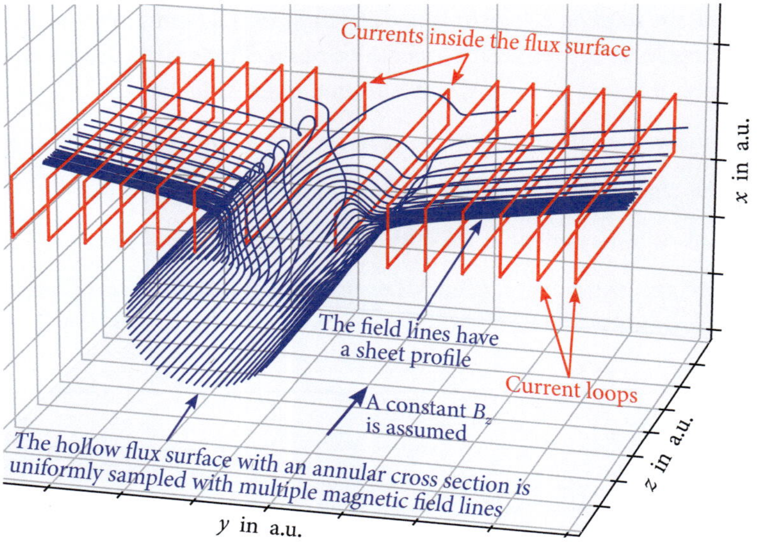

Fig. 3.6: Example for the transformation of the magnetic field lines from an annular cross-section profile to a profile of two sheets

Such a beam-shape transformation is nontrivial. No similar device was found in the literature prior to this work. Pagonakis and the author proposed an idea in [83] to “extract” sheet electron beams from the sides of an annular spent electron beam, as shown in the following example.

Electrons moving in a magnetostatic field should follow the magnetic field lines in the absence of any drift or collision. Therefore, the transformation of the electron beam shape is equivalent to the transformation of the magnetic field lines on the flux surface, where the electron trajectories are located.

An example for the creation of such a magnetic field is given in fig. 3.6. In this example, a uniform magnetic field B_z is assumed for the whole region. The carefully designed coils are simplified as loops of currents for a fast numerical evaluation of the magnetic field (see the evaluation of current segments in appendix A.2). These coils create a transverse component $B_\perp$ of magnetic field, which gradually “drains” the field lines from the ring sideward. The total magnetic field is the superposition of the transverse field $B_\perp$ and the background field B_z. Beam electrons are expected to follow the field lines to be transformed to two sheet beams.

3.3.2 Conceptual Collector Design and Basic Examinations

To numerically verify the principle, a conceptual design is created and investigated using CST Particle Studio with the particle-tracking method. Realistic coil models are used to replace the ideal current segments. The coils are shown in fig. 3.7a. A homogeneous background magnetic field of about 22 mT is extended from the gyrotron magnet by the solenoids. The "rectangular" coils on both sides add a transverse component of about 10 mT to each direction. There is a pair of coaxially inserted coils, like in fig. 3.6 (invisible in fig. 3.7 as it is covered by the other structures). An annular electron beam is injected at the end of the cavity with a beam radius of 10 mm. The beam travels through a conic region (see figs. 3.7a and 3.7b). That conic region has the same length as a real mirror box; however, the diameter and details of a real mirror box are not modeled in order to reduce the complexity of the structure. In this fake "mirror box", the transition of magnetic field should be adiabatic. Otherwise, the electrons would gain large transverse velocities and the beam before the transformation would be thicker than via an adiabatic demagnetization. A thick annular beam is not optimal for the beam-shape transformation in the next step.

The transformed sheet electron beams enter the "box" on each side. In the box there are electrodes parallel to the gyrotron axis. To check the trajectories of the primary electrons, the simulations do not involve secondary electrons, yet. The primary electrons exposed in the drift are collected on the far side (defined in fig. 1.20) of each electrode, except at the last stage. A projected view of the trajectories is shown in fig. 3.7c. When an electron approaches the electrode, whose depression voltage matches the initial energy of that electron, the velocity of the electron diminishes significantly (the color in the figure becomes dark blue). Then, the drift velocity takes over the control of the trajectory. This is a sign for the proper operation of such an MDC design.

A collector efficiency of 86 % is given by the CST simulations for this non-optimized conceptual seven-stage collector design.

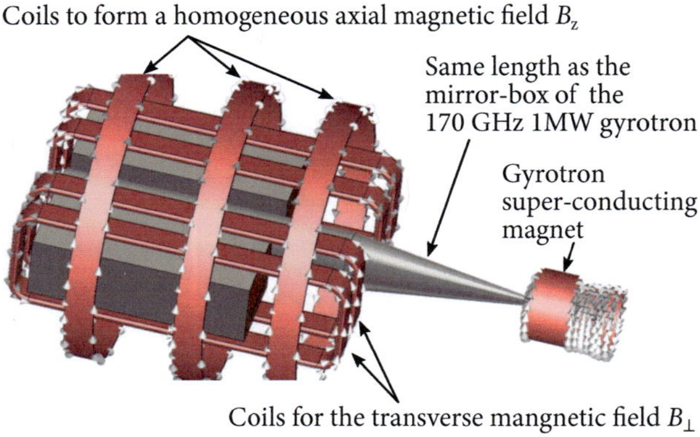

(a) Collector coils

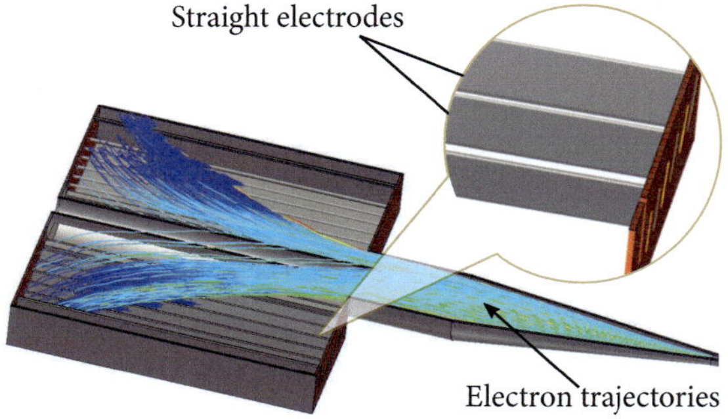

(b) Perspective view of the electron trajectories

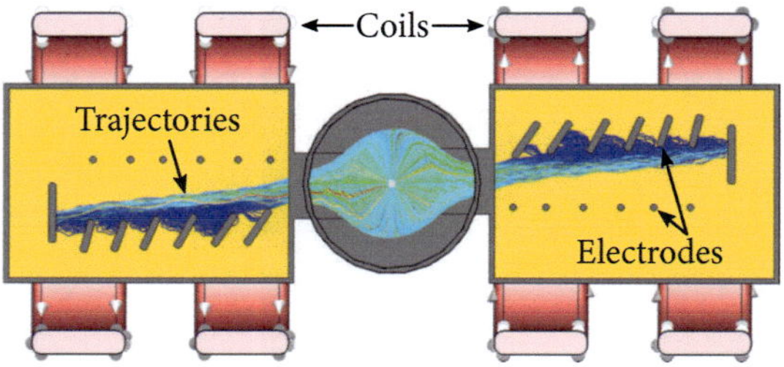

(c) Electron trajectories projected in a cross-section view

Fig. 3.7: CST simulation to verify the concept of sheet beam transformation [85]

3.3.3 Optimization of Electromagnetic Coils Based on the Magnetic Flux Coordinates

The thickness of the sheet beam is an important parameter for the sorting using the planar $\boldsymbol{E}\times\boldsymbol{B}$ drift. An ultra-thin sheet electron beam is ideal for the analysis and will be optimally collected [83]. The beam thickness is determined by two factors:

1. the original thickness of the annular electron beam before the transformation;
2. the magnetic field applied for the beam-shape transformation, which is given by the coil parameters (shape, position, current turns, etc.).

To target the first point, demagnetization before the collector should be adiabatic, as discussed before. For the second point, increasing (compressing) the transverse magnetic field in the box region can reduce the thickness of the sheet electron beam. However, this is not a good option, since the increase of the (total) magnetic field will reduce the effectiveness of the drift, as shown in eqs. (1.15) and (3.4). A general method will be developed in the following paragraphs, in order to facilitate the optimization of the coil parameters.

Green's Function for the Sheet-Beam Transformation System Based on the Clebsch Representation of the Transverse Magnetic Field

To understand the relation between the magnetic lines (flux surfaces) and the current distribution in a cross section of the beam-shape transformation in order to automatically optimize the coil parameters without numerically tracking of the field lines, one has to use the Clebsch representation of the magnetic field. The Clebsch transformation is a mathematical tool originally invented for the hydrodynamics [93]. In the Clebsch transformation of the hydrodynamic equations, two position-dependent scalars α and β form the coordinates of the vortex-lines [94]. This principle was later introduced by Kruskal et al. [95] to describe the coordinates of magnetic field lines in a magnetically confined plasma, which is known as the Clebsch representation of magnetic field:

$$\boldsymbol{B} = \boldsymbol{\nabla}\alpha \times \boldsymbol{\nabla}\beta \, . \tag{3.9}$$

A brief derivation [96] of the Clebsch representation is given as follows. The magnetic vector potential $\boldsymbol{A}$ can be expressed using two scalars (not compatible with the Coulomb gauge):

$$\boldsymbol{A} = \alpha \boldsymbol{\nabla}\beta \,. \tag{3.10}$$

The magnetic flux density is

$$\boldsymbol{B} = \boldsymbol{\nabla} \times \boldsymbol{A} = \boldsymbol{\nabla} \times (\alpha \boldsymbol{\nabla}\beta) = (\boldsymbol{\nabla}\alpha \times \boldsymbol{\nabla}\beta) + \alpha\underbrace{(\boldsymbol{\nabla} \times \boldsymbol{\nabla}\beta)}_{=0}$$

which results in eq. (3.9). The positions in the space with the same value of α (or β) are on the same iso-α (or iso-β) surface. A flux line is the intersection of two surfaces, which means that both α and β are constant along a magnetic field line.

Above is a general introduction for the Clebsch representation of the magnetic field. The following paragraphs present the application of that tool in the sheet-beam transformation.

The axial field component B_z is firstly excluded in the analysis of the coils for the transformation, as B_z can be later linearly superposed to the $B_\perp$ field. Since the coils shown in figs. 3.6 and 3.7 are elongated in $\boldsymbol{z}$ direction, for a simplified model it can be assumed that the coils span infinitely in $\boldsymbol{z}$ direction such that all coils include only the currents, which are parallel to $\boldsymbol{z}$, i.e. $J_x = J_y = 0$ (it will be proven later, that there should also be $\partial J/\partial z = 0$ as a consequence of the assumption). Under this assumption, every cross section perpendicular to the $\boldsymbol{z}$-axis has the identical distribution of the coil current density and thus the identical transverse magnetic field. Such a magnetic field for the sheet-beam transformation has the iso-α and iso-β surfaces depicted in fig. 3.8, which will be quantitatively modeled in the following derivation.

There is more than one possibility to specify α and β in this transformation. One can let β independent of x and y so that

$$\frac{\partial \beta}{\partial x} = \frac{\partial \beta}{\partial y} = 0 \,. \tag{3.11}$$

In order to have iso-α surfaces like in fig. 3.8, it is sufficient to assume for α that

$$\frac{\partial^2 \alpha}{\partial x\, \partial z} = \frac{\partial^2 \alpha}{\partial y\, \partial z} = 0 \,. \tag{3.12}$$

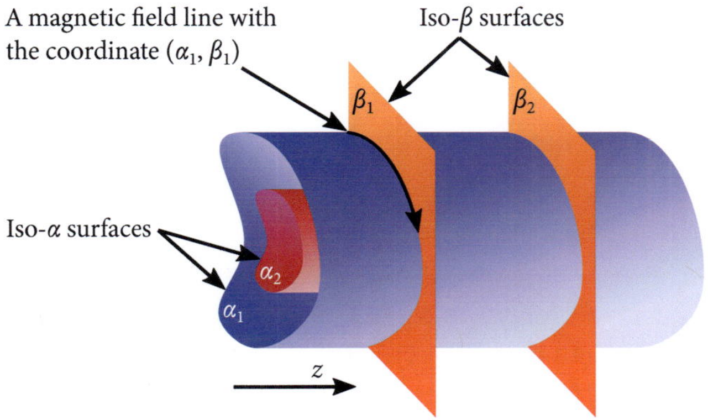

Fig. 3.8: Schematic of the iso-α and iso-β surfaces for the modeling of the beam-shape transformation (without B_z)

In absence of displacement current

$$\boldsymbol{\nabla} \times \boldsymbol{B} = \mu \boldsymbol{J} \,, \tag{3.13}$$

which is

$$\mu \boldsymbol{J} = \boldsymbol{\nabla} \times \begin{vmatrix} \hat{\boldsymbol{x}} & \hat{\boldsymbol{y}} & \hat{\boldsymbol{z}} \\ \frac{\partial \alpha}{\partial x} & \frac{\partial \alpha}{\partial y} & \frac{\partial \alpha}{\partial z} \\ 0 & 0 & \frac{\partial \beta}{\partial z} \end{vmatrix} = \boldsymbol{\nabla} \times \begin{pmatrix} \frac{\partial \alpha}{\partial y}\frac{\partial \beta}{\partial z} \\ -\frac{\partial \alpha}{\partial x}\frac{\partial \beta}{\partial z} \\ 0 \end{pmatrix} = \begin{vmatrix} \hat{\boldsymbol{x}} & \hat{\boldsymbol{y}} & \hat{\boldsymbol{z}} \\ \frac{\partial}{\partial x} & \frac{\partial}{\partial y} & \frac{\partial}{\partial z} \\ \frac{\partial \alpha}{\partial y}\frac{\partial \beta}{\partial z} & -\frac{\partial \alpha}{\partial x}\frac{\partial \beta}{\partial z} & 0 \end{vmatrix}$$

$$\mu \boldsymbol{J} = \begin{pmatrix} 0 \\ 0 \\ \mu J \end{pmatrix} = \begin{pmatrix} \cancel{\frac{\partial^2 \alpha}{\partial x \partial z}\frac{\partial \beta}{\partial z}} + \frac{\partial \alpha}{\partial x}\frac{\partial^2 \beta}{\partial z^2} \\ \cancel{\frac{\partial^2 \alpha}{\partial y \partial z}\frac{\partial \beta}{\partial z}} + \frac{\partial \alpha}{\partial y}\frac{\partial^2 \beta}{\partial z^2} \\ -\left(\frac{\partial^2 \alpha}{\partial x^2}\frac{\partial \beta}{\partial z} + \frac{\partial^2 \alpha}{\partial y^2}\frac{\partial \beta}{\partial z}\right) \end{pmatrix} . \tag{3.14}$$

The terms are canceled because of the assumption in eq. (3.12). The terms next to the canceled ones should be zero

$$\frac{\partial \alpha}{\partial x}\frac{\partial^2 \beta}{\partial z^2} = \frac{\partial \alpha}{\partial y}\frac{\partial^2 \beta}{\partial z^2} = 0 \,,$$

this happens only if their common term $\partial^2 \beta / \partial z^2 = 0$. It means that β has the form

$$\beta = K_1 z + K_0 \,. \tag{3.15}$$

Both coefficients K_0 and K_1 are free to choose. Let $K_1 = -\mu$ and $K_0 = 0$, one gets

$$\beta = -\mu z \,, \tag{3.16}$$

then eq. (3.14) remains

$$\frac{\partial^2 \alpha}{\partial x^2} + \frac{\partial^2 \alpha}{\partial y^2} = J \,. \tag{3.17}$$

This would be a two-dimensional Poisson equation, if one could constrain that J is independent of z (only $J_x = 0$ and $J_y = 0$ are assumed). After applying the derivative of z on both sides of eq. (3.17), the left-hand side of this equation vanishes because of eq. (3.12). In other words, as a consequence of the assumption, the current density must be independent of z:

$$\frac{\partial J}{\partial z} = 0 \,. \tag{3.18}$$

Finally, the relation of the current density J and the so-called Euler potential [97] α is the two-dimensional Poisson equation

$$\nabla^2 \alpha = J \,. \tag{3.19}$$

The value of α at any position $\boldsymbol{r}$ of a cross section can be calculated via the integral of the Green's function for the two-dimensional Poisson equation, which is the logarithm of the distance [98]

$$\alpha(\boldsymbol{r}) = \frac{1}{2\pi} \int J(\boldsymbol{r}') \log|\boldsymbol{r} - \boldsymbol{r}'| \, \mathrm{d}\boldsymbol{r}' \tag{3.20}$$

over the whole cross section. The values of α at the ring of the input electron beam are known. Evaluating α only at the output positions (where the beam should already be transformed to sheets) results in the thickness and the position of the transformed sheet beams, as will be explained in the example later.

Developing a Coil System to Create Ultra-Thin Sheet Electron Beams

A set of ideal coils for a high-quality beam-shape transition is developed as follows. To simplify the evaluation of α values, current segments with zero wire radius like in fig. 3.6 will be used. In that case, the value of α is

$$\alpha(\boldsymbol{r}) = \frac{1}{2\pi} \sum_n I_n \log|\boldsymbol{r} - \boldsymbol{r}_n| \,, \tag{3.21}$$

where I_n and $\boldsymbol{r}_n$ are the current and the position of the n'th current segment, respectively. In a realistic design, these ideal current segments will be replaced with segments of realistic coils, which will have a reasonable thickness and a finite length.

In the following analysis, the transverse projection of the magnetic field will be presented. The projections of the iso-α surfaces shown in fig. 3.9 are equivalent to the projected magnetic field lines. Therefore, they are also called iso-α lines or just magnetic field lines in the following text. Scaling the currents of every coil altogether means to scale the value of α at every position, however, that does not change the shapes of these iso-α lines. In addition, the coordinate system can also be scaled without changing the shapes of the iso-α lines, therefore, the x and y coordinates in fig. 3.9 are relative.

Figures 3.9a to 3.9d demonstrate the steps for the creation of an optimum magnetic field. In fig. 3.9a, four coils (eight current segments) are arranged as a quadrupole. They have the same absolute value of current, whereas the signs of currents are different as shown in fig. 3.9a. Electrons move in the forward direction of the field lines, because the gyrotron super-conducting magnet creates a positive B_z. Two field lines with α of α_1 and α_2 are marked in blue color. They are the reference lines across all figures. The annular beam (the orange ring in fig. 3.9a) will be squeezed by the magnetic field. The right half ring is squeezed to a sheet beam moving rightwards, whereas the left half ring move to the opposite direction. The resulted sheet beam can be analogized to a "sandwich": the electrons originally started at position ① will follow the blue field lines to the middle layer of the "sandwich", while those ones originally started at ② will follow the next outer green line to become one of the outer "bread" for this sandwich-like sheet beam.

Scaling the current of the quadrupole cannot change the shapes of these magnetic field lines. To avoid that sandwich shape, two additional sets of coils are required.

1. Two pairs of coils are inserted coaxially inside the annular beam. The effect of these inserted coils is shown in fig. 3.9b. These coils "push" the blue field lines passing through ① away from the axis, such that those field lines can be closer to the position ② than in fig. 3.9a.

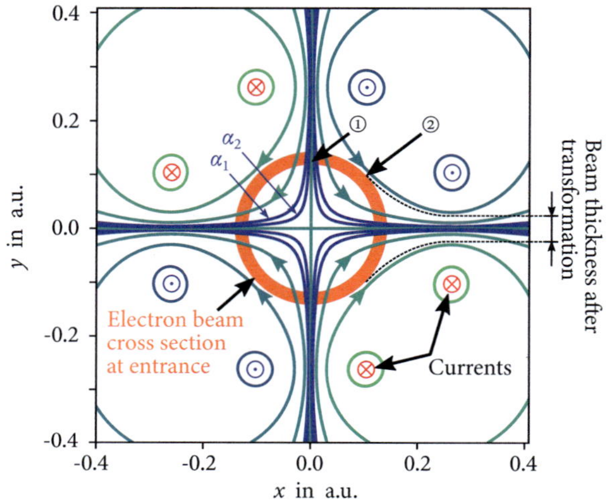

(a) Magnetic field lines of a quadrupole

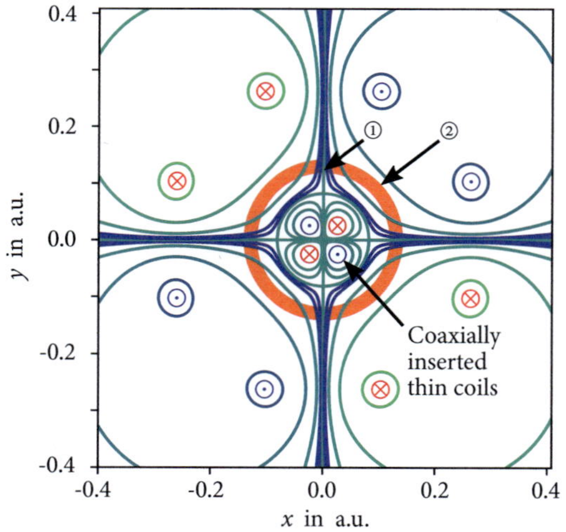

(b) Coaxially inserted coils pushing the field lines

Fig. 3.9: Optimization of the sheet-beam transformation by manipulating the shapes of the iso-α surfaces

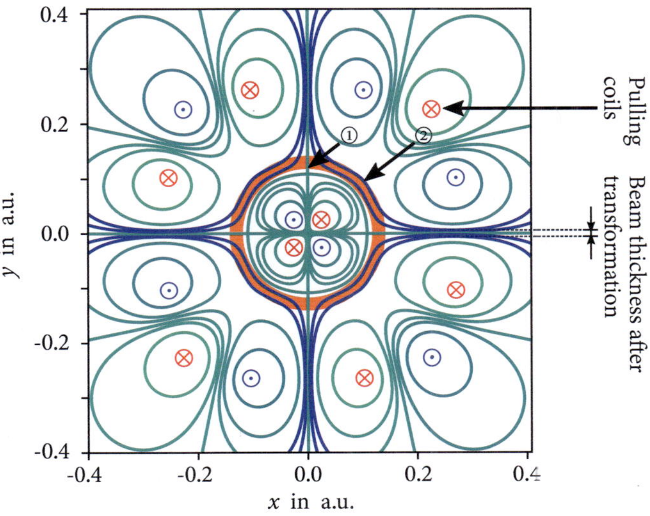

(c) Additional outer coils pulling the field lines

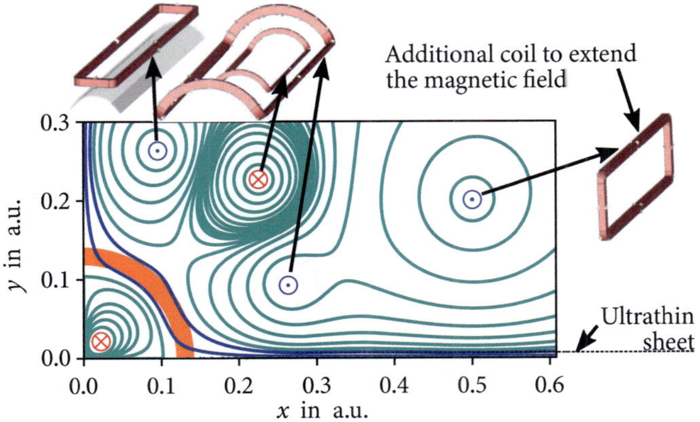

(d) Final magnetic system (first quadrant only) with realistic coil designs

Fig. 3.9: Optimization of the sheet-beam transformation by manipulating the shapes of the iso-α surfaces (continued)

2. From the outer side there should be another pair of coils, which "pull" the blue field lines away from the axis, as shown in fig. 3.9c.

The resulted magnetic field fits better to the annular profile of the beam than only using the quadrupole, such that the thickness of the output sheet electron beam can be significantly reduced.

In order to prevent the sheet beam from diverging, another set of coils around both "boxes" should be placed far away from the center. These side coils extend the magnetic field on the $\boldsymbol{x}$ direction such that there will be sufficient space for the electrodes. The final coil system is shown in fig. 3.9d. There, realistic coils are used to replace the ideal current segments. The center of the MDC geometry is a cylinder (see fig. 3.7c), which has a larger radius than the input spent electron beam. The cylindrical envelope should not intersect the coils, therefore, the transverse current segments of some coils should be bent around this cylinder, as visualized in fig. 3.9d.

3.3.4 Upgraded Designs of Coils and Electrodes

An upgraded design with optimized coils and improved electrodes is simulated for the assessment of the critical parameters. The new magnetic field for the sheet-beam transformation can theoretically produce a sheet electron beam which is three to five times thinner than the one in fig. 3.7. Some special aspects which are different from fig. 3.7 are present as follows.

The first unexpected issue in the coils design is that, as such bent coils from fig. 3.9d are applied to the MDC, their transverse current (the arc segments) can create an unwanted magnetic field. If those coils are close to the electron beam, the magnetic field becomes non-adiabatic. This is similar to the principle in chapter 2. The non-adiabatic magnetic field causes the electrons beam to have a radial oscillation, such that the annular electron beam becomes thicker than expected at the begin of the transformation. Hence, the quality of the sheet-beam transformation decreases, and the output sheet beams become thicker, too. This can be observed in simulations as shown in fig. 3.10a.

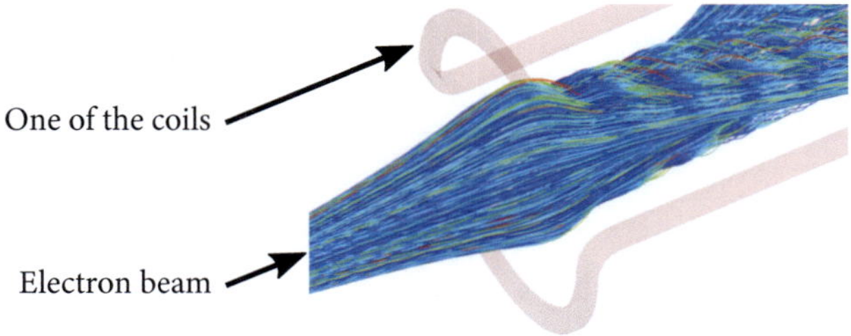

(a) Issue

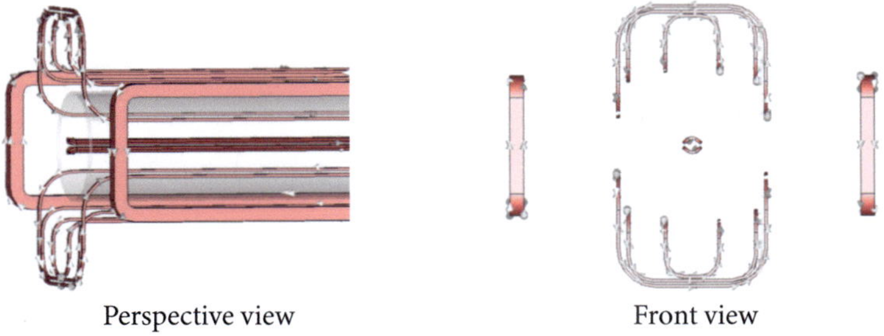

(b) Solution

Fig. 3.10: Upgraded transformation coils to avoid the unwanted non-adiabatic magnetic field produced by the transverse current

The solution is to keep an enough distance between the transverse current segments and the annular beam, as shown in fig. 3.10b. Since the magnetic field is inversely proportional to the distance between the evaluated position and the location of the current, the unwanted magnetic field becomes negligible compared to the axial field B_z, if the problematic current segments are far away from the beam.

There is no physical difficulty to extend the number of collector stages. Therefore, MDC designs with two to five stages are investigated via simulations. The shapes of electrodes are improved as shown in fig. 3.11. Every electrode except the last one in this improved design has a pad at the end. The purpose of these pads is to capture the secondary electrons, which otherwise may escape from the other side of the electrodes due to the drift.

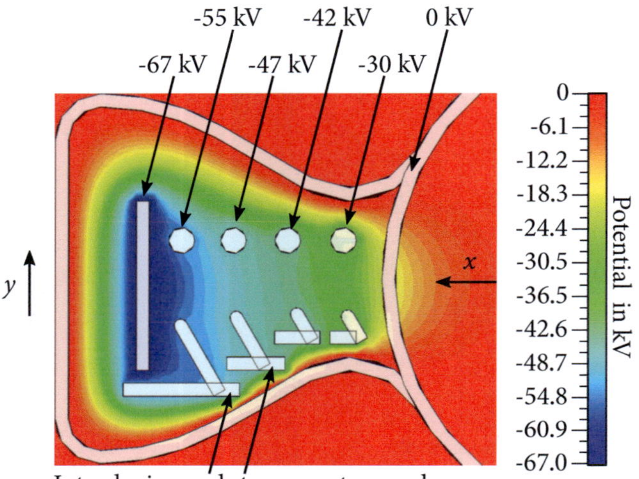

Fig. 3.11: Upgraded electrodes for the sheet beam collector

3.3.5 Assessment of the Critical Parameters

Achievable Collector Size and Mechanical Complexity

The MDC design for the 170 GHz 1 MW gyrotron is investigated via simulations. The length of the MDC is related to the quality of the beam transformation, and the width is related to the energy sorting using **E**×**B** drift.

In order to properly transform the electron beam, the MDC should be at least 30 % (40 cm) longer than the reference SDC. The collector length consists of two parts: the length (distance) for the adiabatic demagnetization and the one for the sheet beam transformation.

- The distance for the adiabatic demagnetization before the transformation depends on (1) the field profile of the gyrotron main SC magnet, (2) the original beam radius after the interaction and (3) the input beam radius of the beam-shape transformation. The demagnetization may need a longer length than the length of the mirror box (approximately 0.5 m). In order to guarantee the adiabaticity and avoid the unnecessary increment of beam thickness, this length can hardly be reduced.

- The length of the sheet-beam transformation can be varied by scaling the current of the transformation coils. High currents produce a high transverse magnetic field. The high magnetic field shrinks the distance required for the transformation; however, the drift is reduced for two reasons. First, as $B_\perp$ increases, the angle ϑ between electric and magnetic field becomes smaller. Second, if the magnitude of the total magnetic field increases, both drift velocity $\boldsymbol{v}_\mathrm{d}$ and distance D will decrease. Hence, it is also not easy to shrink the distance of transformation.

Another possibility to shrink the length for the transformation is to increase the number of "boxes" as shown in fig. 3.12. For instance, transforming the annular beam to three sheet beams will shrink the length of the previous MDC design proposal by one third. However, more coils and electrodes are required and therefore, the MDC becomes more complicated.

The sheet-beam transformation requires a typical radius of 300 mm for the quadrupole coils. Each "box" has a depth of 100 mm to 200 mm, depending on the number of collector stages. Hence, the total width of an MDC with two "boxes" exceeds 1 m, which is twice of the diameter of the reference SDC.

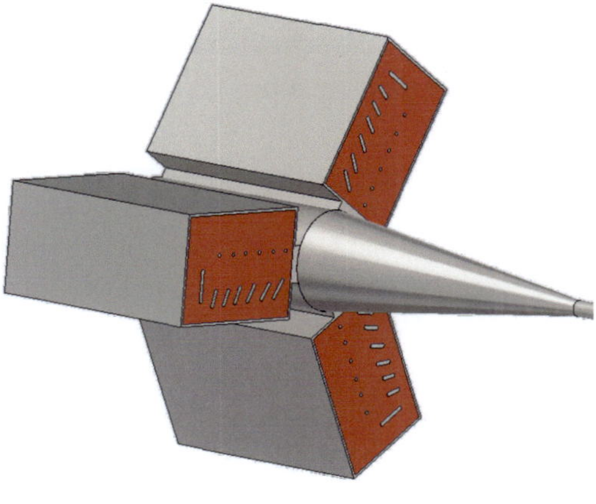

Fig. 3.12: Triple boxes make the collector shorter, however, more complex

Totally 10 coils and up to 4 different numbers of current · turns (not including the coils for the homogeneous B_z field) are required for an optimized MDC of this type. That is more complicated than the existing sweeping coil system (six for the transverse sweeping [99] and one for the vertical sweeping) of an SDC. In addition, the shapes of the coils here are different from the regular solenoids. On the other hand, increasing the number of collector stages is theoretically straightforward.

Expected Collector Efficiency and Unwanted Back-Stream Current

The spent electron beam of the 170 GHz 1 MW gyrotron with the energy spectrum shown in fig. 1.9 are considered in the simulations. A two-stage collector with depression voltages of 30 kV and 45 kV can achieve a collector efficiency of 80.6 % in the simulation, if secondary electrons are not considered. This efficiency is very close to the theoretical optimum, which would be 82.4 % according to table 1.2. The five-stage depressed collector shown in fig. 3.11 should have 86 % collector efficiency for a realistic electron beam, also in the absence of secondary electrons. A simulation with the CST built-in Furman emission model [36] of copper shows that the collector efficiency of the five-stages collector is reduced to 82 % if secondary electrons are taken into account. The simulations also show that the most problematic secondary electrons are emitted from the last stage. An optimization on the shapes of electrodes could further reduce the influence of secondary electrons. It is observed in the simulation, that neither primary nor secondary electrons are back-streamed.

Maximal Expected Power Load Density on the Critical Parts

Since the electrodes are straight stabs, the channels for the cooling liquid can be straight, too. This could be an advantage for the engineering design. However, the current of the transformed sheet electron beam is not uniformly distributed. The density of the sampled magnetic field lines in fig. 3.6 clearly shows that there are more beam electrons (thus higher current) transformed to the front (mirror-box) end of the sheet beam than to the rear end. Hence, a hotspot can be seen in fig. 3.13, where the load is higher than 1 kW/cm^2 in the simulations of a multistage collector design for the 170 GHz 1 MW gyrotron. The load at the other parts is lower than 400 W/cm^2, i.e. in the acceptable range. For the future 2 MW DEMO gyrotron, the power load density is approximately doubled, such that except this hotspot, some other places could also exceed the given 500 W/cm^2 limitation.

There are several proposals to decrease the power density:

- optimizing the angle of the electrodes (adjusting the tilt in the x-y plane),
- tuning the depression voltages (but leaving the efficiency non-optimum),
- increasing the number of stages,
- increasing the number of "boxes" to split the total power.

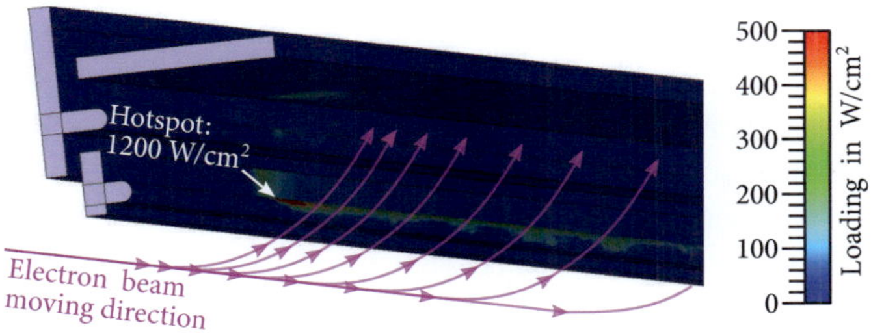

Fig. 3.13: Typical thermal load at the last three stages of a sheet-beam MDC

Summary

The major advantages of this MDC proposal compared to other $\boldsymbol{E}\times\boldsymbol{B}$ MDCs, which will be introduced in the next sections, are

- simple shapes of electrodes,
- good extendibility for the number of stages (a higher number of stages results in a better efficiency and the lower power load density)

The drawbacks are obviously:

- large size (in both length and width),
- complicated additional coils.

Therefore, this may not be the best MDC proposal for the 170 GHz 1 MW gyrotron, nor for the future DEMO gyrotron. The drawbacks are mainly caused by the complexity of the sheet-beam transformation and the required size for that. To target these drawbacks, extremely elongated elliptic beam and arc beams can also be handled as generalized sheet beams. MDC proposals for the collecting of the generalized sheet beams will presented in sections 3.4 and 3.5.

3.4 Proposal II: Planar Drift Applied to an Elongated Elliptic Electron Beam

3.4.1 Creation of an Elongated Elliptic Electron Beam

An elliptic electron beam has a semi-major axis a and a semi-minor axis b as shown in fig. 3.14. An extremely elongated electron beam (when $a \gg b$) can be sorted and collected using the $\boldsymbol{E}\times\boldsymbol{B}$ drift as a (generalized) sheet electron beam, which has been mentioned in section 3.3. The method to transform the gyrotron spent electron beam to an elongated elliptic beam is presented in this section.

An elongated elliptic electron beam can be obtained from the annular beam using the combination of two ideal coils shown in fig. 3.15. These coils are arranged in a straight form of the Yin-Yang coil [100], the latter was proposed for the magnetic confinement of the fusion plasmas. Here in fig. 3.15:

- the front (i.e. left) coil defocuses the injected beam in y-direction,
- the rear coil compresses the beam cross section in the other direction.

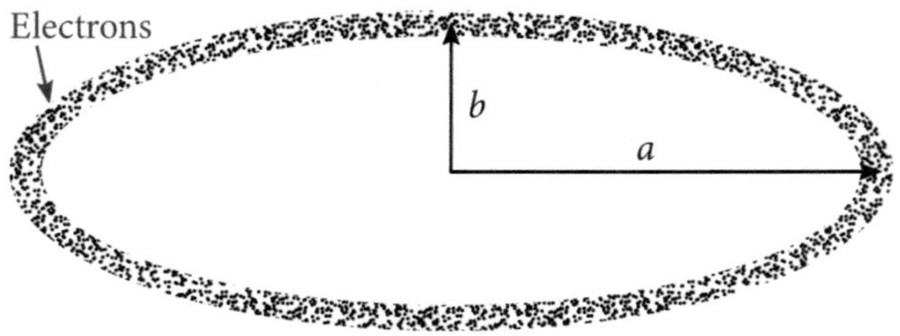

Fig. 3.14: Cross section of an elongated elliptic electron beam

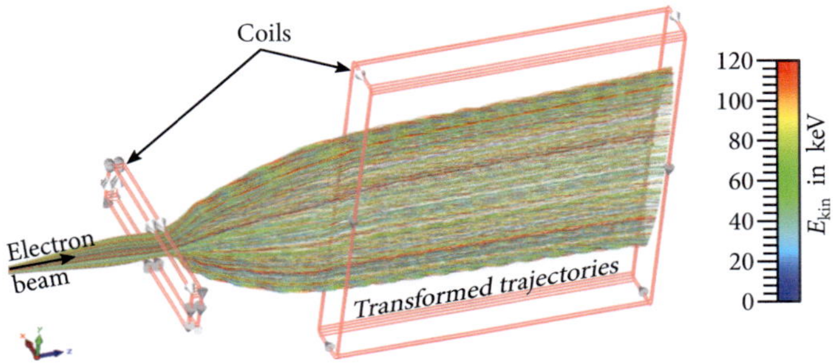

Fig. 3.15: Transformation from annular electron beam to elongated elliptic beam using ideal coils [86]

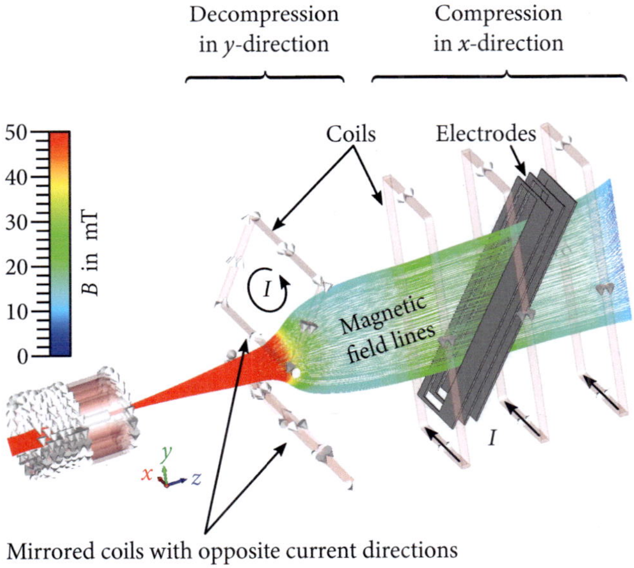

Fig. 3.16: Elongated elliptic beam transformation using more reasonable coils [86]

The simulation of the electron trajectories in fig. 3.15 also takes into account the stray magnetic field of the gyrotron main SC magnet.

The ideal coils can be realized as follows. The front coil is replaced by two coils, so that the electron beam does not need to penetrate the coil loop, as visualized in fig. 3.16. The rear coil is split into three rectangular coils. The rectangular coils have a wide aperture, in order to place electrodes inside the loops. The electrodes shown in fig. 3.16 can be constructed and isolated in a similar way like the TEF collector [61], see fig. 1.17.

3.4.2 Estimation of the Minimal Electrode Length

In order to estimate the length of an elliptic-beam MDC, the axial length of an electrode as a part of the total length, is estimated in this section. The basic requirement for the collection of all electrons using a drift is that the drift distance D (see fig. 1.20) should be longer than the thickness of the electron beam. The elliptic electron beam has a thickness of $2b$:

$$\frac{D}{2} > b\,.$$

Substituting D by eq. (3.4) there is

$$\frac{\min\{v_0\}\, m}{B\, q} \tan\vartheta > b \tag{3.22}$$

where v_0 is the axial velocity of an electron at the entrance of the **E**×**B** region. Obviously, the elliptic electron beam should enclose the same magnetic flux ψ as the original annular electron beam:

$$\psi = \pi\, B_0\, {r_0}^2 = \pi\, B\, a\, b\,, \tag{3.23}$$

where B_0 and r_0 are the initial magnetic flux density and beam radius of the annular electron beam, respectively. In this way, the flux density B in eq. (3.22) can be expressed as a function of the beam-enclosed magnetic flux ψ. The other variable required in eq. (3.22) is the angle ϑ in fig. 3.17, which is

$$\tan\vartheta = \frac{L}{2a}\,, \tag{3.24}$$

where L is the longitudinal axial length of one electrode. Substituting eq. (3.23) and eq. (3.24) into eq. (3.22) yields the minimal axial length of an electrode

$$L_{\text{min}} = \frac{2\psi q}{\pi m \min\{v_0\}} . \tag{3.25}$$

The length of the $\boldsymbol{E}\times\boldsymbol{B}$ region is the axial length of a single electrode (L_{min}) plus the total distance between the electrodes and the axial thickness of all electrodes. The total collector length is the sum of the required length for the $\boldsymbol{E}\times\boldsymbol{B}$ region and the axial length for the elliptic-beam transformation.

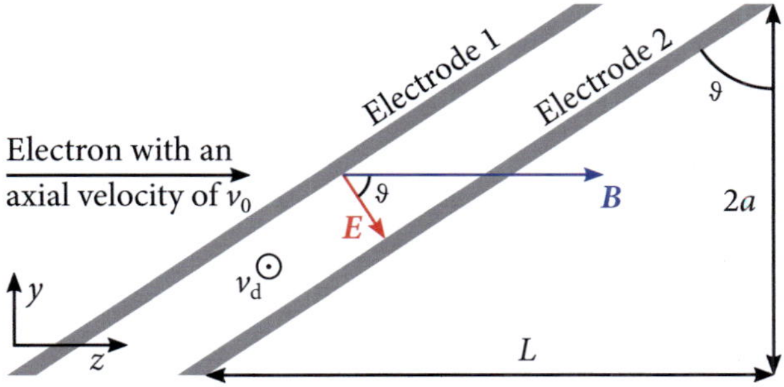

Fig. 3.17: A side view of the electric field, magnetic field and the electrodes for an elliptic-beam MDC

3.4.3 Assessment of the Critical Parameters

Achievable Collector Size and Mechanical Complexity

Equation (3.25) shows that the axial length of an electrode L_{min} does not depend on the ratio of the ellipse axes, but only on the magnetic flux enclosed inside the elliptic beam. The enclosed flux is determined by two parameters: the cavity flux density B_z, which is related to the RF frequency, and the enclosed area at the corresponding axial position. Here are some gyrotron examples for the theoretical estimation of the minimal electrode length:

- The 28 GHz, 10 kW second-harmonic gyrotron [101] corresponds to an L_{min} less than 6 cm. If the transformation also takes place in a short distance, then such an MDC for this gyrotron can be very compact.

- According to the magnetic flux and the electron energy in [102], the 74.2 GHz, 100 kW gyrotron should have $L_{\text{min}} \approx 70$ cm for $\min\{v_0\} \hat{=} 7$ keV. The elliptic beam approach may still be applied.
- The Russian ITER gyrotron [103] has a smaller beam radius than the EU variant. Hence, it has $L_{\text{min}} \approx 1.2$ m.
- The EU ITER $TE_{32,9}$ hollow-cavity gyrotron [24, 26] (i.e. the reference 170 GHz 1 MW gyrotron) has a large magnetic flux enclosed in the electron beam, therefore it would have $L_{\text{min}} \approx 2$ m.
- The 170 GHz 2 MW $TE_{34,19}$ coaxial-cavity gyrotron [104] has a magnetic flux, which is approximately 20 % higher than the $TE_{32,9}$ hollow-cavity gyrotron. Its L_{min} should also be 20 % longer.
- For a DEMO gyrotron with a higher frequency, e.g. 204 GHz, an even higher magnetic flux is possible and such an MDC might have a further 20 % longer L_{min}.

This is only the axial length of one electrode. The total length of the MDC should be even larger, since a sheet-to-elliptic beam transformation is required. For the last three examples, this concept will result in a very large collector size, which might be significantly longer than 3 m. Although L_{min} does not depend on the beam thickness $2b$, the collector efficiency indeed is related to the beam thickness [83]. Any attempt to improve the collector efficiency or to increase the drift distance will further increase the collector length.

Summary

This approach might be suitable for low-flux gyrotrons, however, it is not practical for the reference 170 GHz 1 MW gyrotron and also not for the future DEMO gyrotron, because of the resulted large size. Therefore, no realistic design is investigated. The fundamental reason for the large size is the conservation of magnetic flux enclosed in the small-orbit electron beam.

3.5 Proposal III: Planar Drift Applied to Arc Electron Beams

3.5.1 Creation of Arc Electron Beams

As the applicability of proposal using elongated elliptic electron beam is limited by the enclosed magnetic field, another proposal is presented in this section to break the topology of the electron beam, in order to circumvent that limitation.

The idea can be explained with fig. 3.18. Based on the previous proposal, the first simulation model for the demonstration of the principle also involves an elliptic-beam transformation like in fig. 3.16. It is checked by the simulations, that the transformed beam in this case can be less elliptic (has a larger b/a ratio) than the one in section 3.4, which means that such a transformation could be easier to achieve. Two baffles shown in fig. 3.18 collect the electrons at both ends of the major axis of the elliptic beam cross section. This takes place before the electron beam enters the drift region of the collector. In this way, the elliptic electron beam is broken into two (elliptic) arc beams. Then, each arc beam is sorted and collected as a curved sheet beam by a planar $\boldsymbol{E}\times\boldsymbol{B}$ drift. The difference to the collection of a regular sheet beam is that, the shapes of the electrodes should match the curvature of the arcs in order to collect these arc beams optimally. The $\boldsymbol{B}$-field in the drift region (right-hand side of fig. 3.18) has only a $\mathbf{z}$-component, whereas the electric field has a perpendicular component which will cause a downward drift in this view.

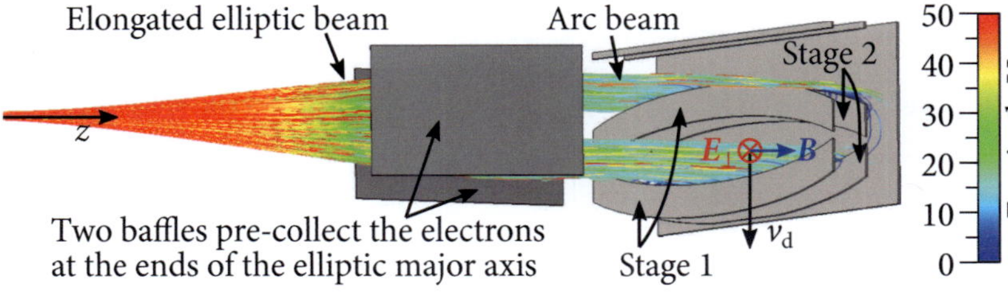

Fig. 3.18: Breaking an elliptic electron beam into two curved sheet beams

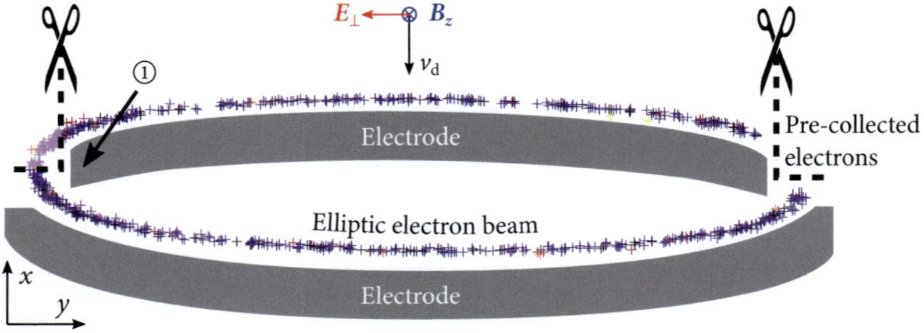

(a) Split an elliptic beam into two elongated arc beams

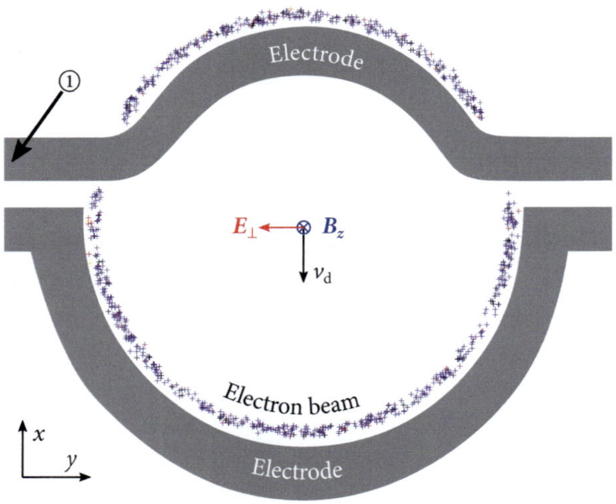

(b) Split an annular beam into two arc beams

Fig. 3.19: Transverse projections of the (initial) arc electron beams and the electrodes [86]

The electrons collected by the baffle are decelerated by the depression voltage of the first stage, no matter how much initial kinetic energy the electrons have. Hence, they are not optimally decelerated. To minimize the number of these electrons, the optimal baffles are not exactly aligned to the ellipse axis, instead, they only need to pre-collect a part of the electron beam at the ends of the major axis (see fig. 3.19a), such that the length of the resulted two arcs are not identical.

To simplify the principle, it was found by simulations that the flattening or ellipticity of the electron beam does not affect the collector efficiency. Therefore, the annular beam can be directly broken into arcs, as shown in fig. 3.19b. That avoids the increase of the collector length for the elliptic-beam transformation. As the pre-collected electrons will not collide with any objects later, position ① in fig. 3.19 can be used to hold and isolate the electrodes like in the construction schematic of fig. 1.17.

Compared to the approach collecting the unbroken elongated elliptic beam in section 3.4, this proposal has three advantages:

1. The beam thickness is reduced from $2b$ to the order of Larmor radius (typically, $r_{\mathrm{L}} < 2\,\mathrm{mm}$). A thinner beam results in a better collector efficiency [83].
2. The shapes of electrodes match the shapes of electron beams. A similar drift distance is required for each electron. Such a drift can be achieved within a reasonable geometric size even for a high-flux gyrotron.
3. The regular arc beams in fig. 3.19b do not need a special beam-shape transformation. Therefore, the collector is shorter than the case using an elliptic electron beam.

3.5.2 Estimation of the Total Collector Length

The analysis of the MDC length will be presented in the following paragraphs. As the length is estimated, the other parameters such as the width, weight, and drift distance are also known [86]. Here, the annular electron beam is considered for the estimation, since it requires a shorter collector length than for an elliptic beam. The total length L_{total} of the $\boldsymbol{E}\times\boldsymbol{B}$ MDC for arc electron beams consists of three parts:

$$L_{\mathrm{total}}(B, D) = L_{\mathrm{trans}} + L_{\mathrm{electrodes}} + \sum L_{\mathrm{gap}} \,. \qquad (3.26)$$

It is a function of two variables: the magnetic flux density B and the desired drift distance D.

The first term L_{trans} on the right-hand side of eq. (3.26) is the length for the transition of the magnetic field prior to the **E**×**B** region. The axisymmetric magnetic field should be smoothly reduced to the required magnitude (in the order of 20–80 mT) of the **E**×**B** collector. The transition should be adiabatic, in order to minimize the expansion of the final beam thickness. The most convenient approach is to include only the stray magnetic field $B(z)$ of the gyrotron main SC magnet without any additional modification to ensure a thin electron beam.

L_{trans} is a function of the final magnetic field B. The key to formulate $L_{\mathrm{trans}}(B)$ is to find the inverse function of $B(z)$. Since the coils are far away from the collector, a proper simplification is to mathematically treat the set of coils in the gyrotron main SC magnet as an ideal current loop. The axial magnetic field of an ideal current loop (see fig. A.1 on page 159) according to the Biot-Savart law is

$$B_z = \mu \frac{I}{2} \frac{r^2}{\sqrt{r^2 + (z - z_0)^2}^3} , \tag{3.27}$$

where r is the loop radius and z_0 is the axial location of the loop center. For $|z - z_0| \gg r$, i.e. in the far field of the loop, eq. (3.27) is approximately

$$B_z \approx \frac{K}{|z - z_0|^3} , \tag{3.28}$$

which has the inverse function

$$z = \left(\frac{K}{B_z} \right)^{1/3} + z_0 . \tag{3.29}$$

Fitting the field data of the 170 GHz 1 MW EU gyrotron to eq. (3.29), K is found to be 0.069 T m^3. The fitted z_0 is the middle point of the main SC coil, as expected. This equivalent current loop model with far-field approximation only has 0.1 % modeling error in the transition region, and it is even accurate enough for off-axis positions at a reasonable radius.

Finally, the transition length L_{trans} is the distance from the end of the mirror box z_{m} to the begin of the **E**×**B** region where the field $B = B_z$ starts to be uniform

$$L_{\mathrm{trans}} = \left(\frac{K}{B} \right)^{1/3} + z_0 - z_{\mathrm{m}} . \tag{3.30}$$

The second term $L_{\text{electrode}}$ on the right-hand side of eq. (3.26) is the axial length of one single electrode. Replacing $2a$ by $2r$ in fig. 3.17, the minimum $L_{\text{electrode}}$ is approximately

$$L_{\text{electrode}} = 2\,r\,\tan\vartheta\,. \tag{3.31}$$

The radius r of the annular beam is a function of the homogeneous magnetic flux density B_z regarding the conservation of magnetic flux (the one derived from Busch's theorem). The term $\tan\vartheta$ is the inverse function of eq. (3.4), which takes B_z and D as parameters. The axial length (height) of a single electrode is then

$$L_{\text{electrode}}(B, D) = 4\frac{q}{m_{\text{e}}}\frac{D}{v_0}\sqrt{\frac{B\,\psi}{\pi}}\,. \tag{3.32}$$

The last term of eq. (3.26) is the sum of the distances between electrodes. Each distance can be an arbitrary value, as far as the electric field does not break down (in this estimation, a distance of 10 cm between the adjacent stages is considered, thus ideally there is an electric field of less than 2 kV/cm between two parallel electrodes). This term is insignificant compared to L_{trans} and $L_{\text{electrode}}$.

Figure 3.20 is the relation between the drift distance D, the homogeneous collector magnetic field $B = B_z$ and the collector length of a three-stage collector. The minimal initial velocity of the electrons entering the collector is approximately $v_0 \approx 8$ keV. Each iso-D curve has a minimum of L_{total}, since

- The right side of the diagram has a high B_z. That means a short distance for the demagnetization. However, to keep the drift distance constant, the tilting angle ϑ of the electrodes must be larger. Therefore, the total length is longer.
- On the left side of the diagram, although the tilting angle of electrodes decreases as B_z diminishes, the distance required for the demagnetization increases non-linearly. Furthermore, the beam radius should also be larger in order to enclose the same magnetic flux. Thus, the total collector length increases dramatically.

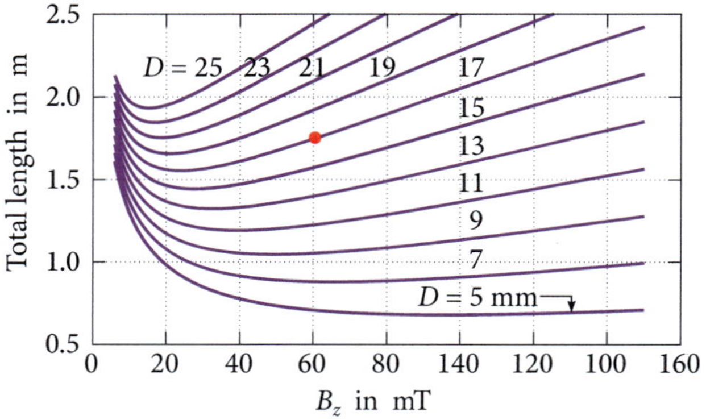

Fig. 3.20: Estimated total length of a three-stage arc-beam collector for a drift distance D from 5 mm to 2.5 cm [86]. The iso-D curves are in violet. The red dot is the operation point for a conservative MDC design in section 3.5.3.

3.5.3 Conceptual Collector Design and Basic Examinations

A conceptual design of the arc-beam collector for the 170 GHz 1 MW gyrotron is created and simulated with CST. It requires four additional collector coils to sustain a homogeneous magnetic field of $B_z = 60$ mT. The reason of choosing this magnetic field is explained as follows. The drift distance D should be higher than the beam thickness, while the latter is up to 5 mm. In this design, D is chosen to be approximately 17 mm. On the curve for $D = 17$ mm in fig. 3.20, the magnetic field for the shortest collector length is about $B_z \approx 30$ mT. However, a more conservative operation point (the red dot in fig. 3.20) with a high magnetic field at $B_z \approx 60$ mT rather than at 30 mT is chosen for four reasons:

- The collector diameter with $B_z = 60$ mT is smaller than that at the operation point of $B_z = 30$ mT. (A smaller diameter does not mean a higher power load density, since the beam impacting area depends also on ϑ.)
- The electron beam at $B_z = 60$ mT is thinner than the one at $B_z = 30$ mT.
- The electron beam has a stronger confinement at $B_z = 60$ mT than at 30 mT; therefore, the MDC operation should be more stable against the external perturbation magnetic field.

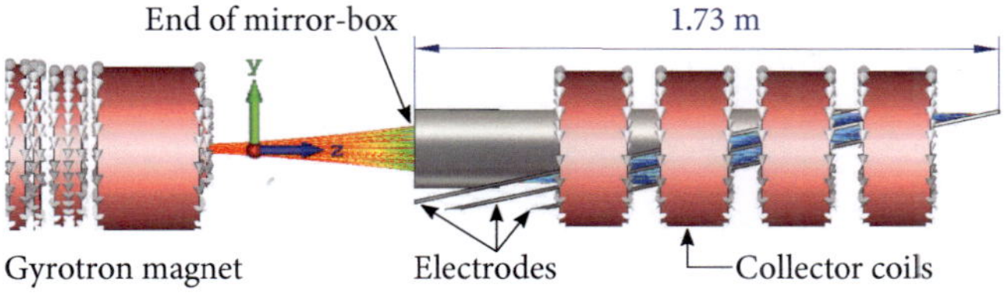

(a) Side view

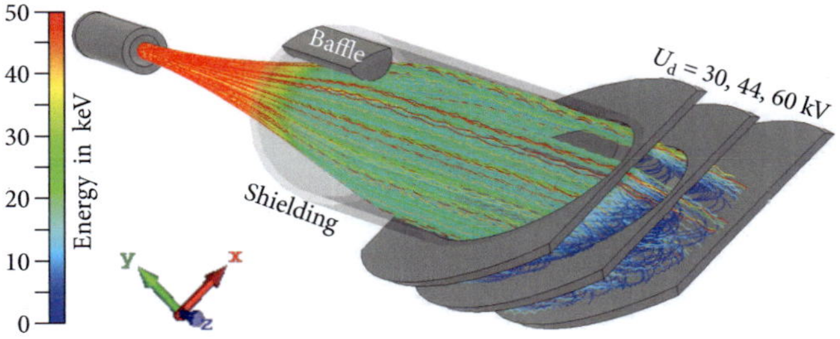

(b) Perspective view (the coils not shown)

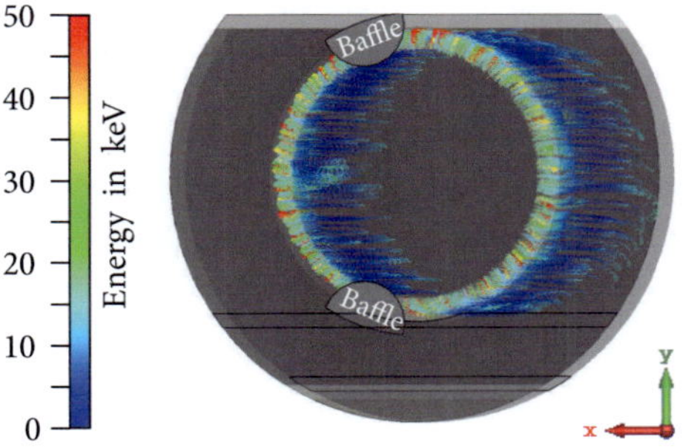

(c) Transverse projections of the electrodes and the electron trajectories. The tops of the disk-form electrodes are not necessary for the creation of the electrode field. Therefore, they are removed for a shorter collector length.

Fig. 3.21: Design of a three-stage arc-beam depressed collector with the simulated electron trajectories [86]

- It is preferred to stay on the conservative side of fig. 3.20, where the collector length does not vary with the magnetic field dramatically.

A side view of the design is shown in fig. 3.21a. Before the ***E***×***B*** drift is applied, the annular electron beam is split into two arcs by a pair of optimized baffles, which are located at the optimum positions for the pre-collection of the electrons at those azimuthal ranges, as shown in figs. 3.21b and 3.21c. The baffles have the same depression voltage as the first stage. Therefore, they are able to recuperate a part of the kinetic energy of those pre-collected electrons. There is a cylindrical wall shielding the electrons from the collector entrance to the first stage. Otherwise there would be an unwanted electric field, which would distort the electron beam before the beam enters the carefully designed ***E***×***B*** drift region.

To verify the principle, a two-stage variant and a three-stage variant are simulated. Both have the same magnetic field and the same angle ϑ. The two-stage variant should have an $\eta_{\text{col}} = 74\,\%$ with $U_{\text{d}} = 30$ and 45 kV, according to the simulation, while the three stages variant with $U_{\text{d}} = 30$, 44 and 60 kV should have an $\eta_{\text{col}} = 82\,\%$. These simulations ignore secondary electrons, but they take into account the space charges of the electron beams.

The transversely projected electrodes have the form as shown in fig. 3.21c. The inner electrodes can be held and isolated at the azimuthal ranges, where the sectors of electron beam are pre-collected. In order to keep the simulation model simple, the inner electrodes of the first two stages are elongated disks instead of curved strips depicted in fig. 3.19b.

3.5.4 Assessment of the Critical Parameters

Achievable Collector Size and Mechanical Complexity

The sample MDC design is 30 % longer than the reference SDC. The MDC has a maximal electrode radius of 250 mm, which is 2.5 cm larger than the reference SDC. Four coils are required to keep the magnetic field in the collector homogeneous. They consume up to 800 W of power (assuming copper wires). This type of MDC can be similarly constructed like the Tilted Electric Field

(TEF) collector, which has been built and experimented in the TWT [61]. In principle, adding more stages to the design is straightforward.

Compared to the designs of the sheet-beam approach and of the elliptic beam approach, the design presented in this section is considered simpler:

- This approach results in a significantly smaller collector size than that of the previous two $\boldsymbol{E}\times\boldsymbol{B}$ MDC proposals.
- The required collector coils are more regular than in the previous two proposals.

Expected Collector Efficiency

Since the energy in the pre-collected electrons cannot be optimally recuperated, the maximal achievable collector efficiency is mainly limited by this energy loss. That means, even an ideal MDC using this approach has a lower collector efficiency than the estimated value in table 1.2. Taking the optimized baffles in fig. 3.21 for example, an ideal MDC using an infinite number of stages and an ideal energy sorting would have 92 % collector efficiency rather than 100 %.

Without secondary electrons, the design of a two-stage collector assuming the spent electron beam spectrum of the 170 GHz 1 MW gyrotron in fig. 1.9 has a theoretical collector efficiency of 74 %. A three-stage collector has $\eta_{\mathrm{col}} = 82\,\%$. These values are evaluated without secondary electrons.

In the simulations considering secondary electrons, the Furman model for copper [36, 37] implemented in CST is applied for the electrodes. A straightforward activation of secondary electron model reduces the collector efficiency of the three-stage collector from 82 % to 70 %. The major reasons for such a large efficiency drop are identified. Two of them can be handled just with minor modifications to the electrodes:

- Secondary electrons emitted at both vertices of the inner electrodes have less chance to be collected again at that stage. They bounce back to the previous stage and reduce the collector efficiency. Rounding the edges like fig. 3.22a can gain 3 % collector efficiency from the secondary electrons.

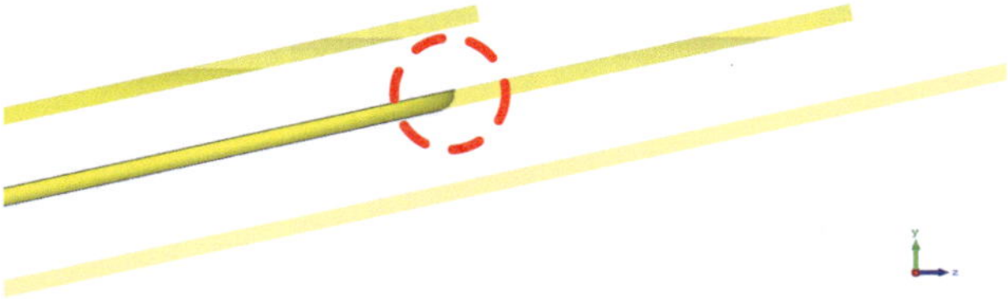

(a) Rounding the inner electrode

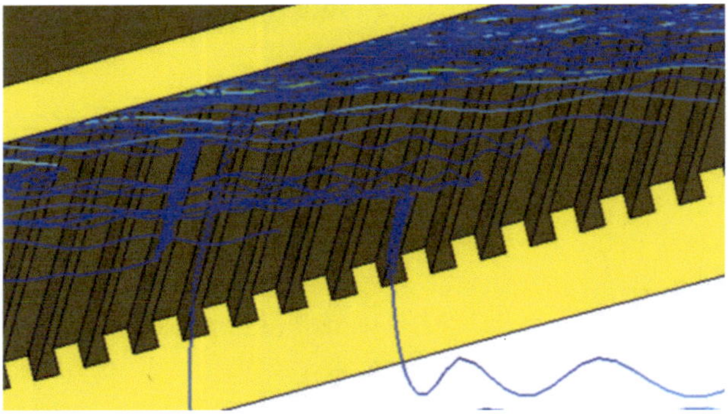

(b) Grooves parallel to drift

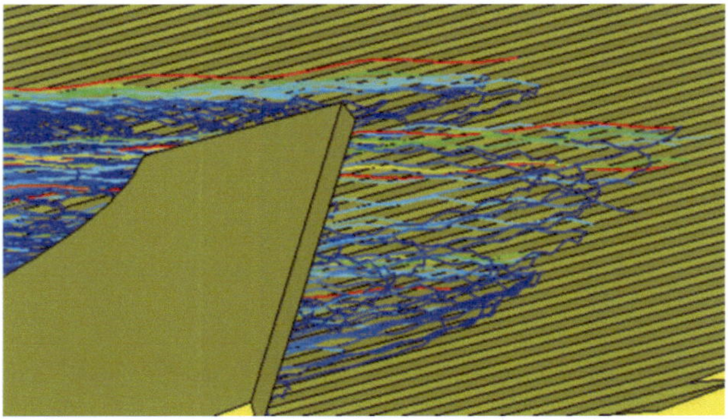

(c) Grooves perpendicular to drift

Fig. 3.22: Suppressing secondary electrons in the arc-beam MDC [86]

Compared to a sharp edge, the rounding increases the area for the re-collection of secondary electrons.

- It is noticed that the most problematic secondary electrons are emitted at the last stage, as they are backwards accelerated by the local electric field. Grooving the surface of the last stage (introduced in section 2.5.2) can affect the effective SEY. Different orientations of numerous 5 mm wide by 5 mm deep grooves are compared in the simulations. In the first case (fig. 3.22b) the grooves are parallel to the drift direction. The secondary electrons are able to move along the grooves and even escape the electrode. In the second case (fig. 3.22c) the grooves are perpendicular to the drift direction. Although the emission of secondary electrons cannot be fully suppressed. The grooves reduce the effective emission ratio of secondary electrons and raise the collector efficiency by additional 3 % points.

Applying these two improvements, the three-stage collector simulations show a collector efficiency of 76.5 % considering secondary electrons. The most effective solution for a further suppression of the secondary electrons is to have enough space at the last stage in order to tune the local electric field and the angle of the magnetic field to the electrode. However, that will increase the complexity of the design.

In the simulations, no current of primary electrons is back-streamed, whereas less than 10 mA of back-stream secondary electron current is observed. This current is in the same order as the back-streamed secondary electron current of the reference SDC and thus the back-streaming can be considered harmless.

Maximal Expected Power Load Density on the Critical Parts

The beam power distributed on the electrodes of this MDC is more uniform than in the case of fig. 3.13. The load on the electrodes is controllable through changing the beam radius and the tilting angle ϑ of electrodes.

However, the baffles collect the beam electrons despite of their kinetic energy with the same depression voltage. Hence, the baffles not only limit the maximally achievable efficiency, but also cause a high power load on them. The critical power load is shown in fig. 3.23. The power load density on everywhere else is below 500 W/cm^2 for the 170 GHz 1 MW gyrotron spent electron beam except at both baffles, where the load can be higher than 3 kW/cm^2. (The most peak value might be numerical artifacts, which is currently hard to process in CST.) Although the power density on the baffles can be reduced by optimizing the electron impacting angles or beam radius, the use of the baffles can result in an unacceptable static power load density for a gyrotron with an output power of 1 MW. For a future DEMO gyrotron which will output 2 MW RF power, the baffles will be more problematic.

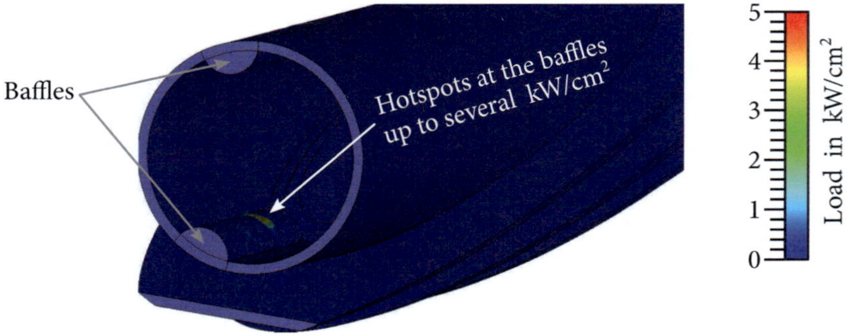

Fig. 3.23: Thermal load at the baffles

Summary

In this section, the annular electron beam is split into two arc beams. An arc beam can be collected as a curved sheet beam. Therefore, this approach is considered as another approach to apply the planar $\boldsymbol{E}\times\boldsymbol{B}$ drift for the energy sorting of sheet electron beams.

The major advantage compared to the other two MDC types using planar $\boldsymbol{E}\times\boldsymbol{B}$ drift in sections 3.3 and 3.4 is the smaller size, which is only 30 % longer than the reference SDC and has a reasonable collector radius.

However, two baffles are needed to split the electron beam. There is a high power load density on them and the energy recuperation at the baffles can never be optimal. Therefore, this kind of MDC might need three stages to achieve the goal for an $\eta_{\text{total}} > 60\,\%$.

In order to realize this idea, better cooling technologies (e.g. the mini-channel cooling [105]) for the local intensive power load on the baffles combining with a larger beam radius might be applied. The possibility to scale this approach for a 2 MW DEMO gyrotron will also be limited by this critical power load.

3.6 Proposal IV: Radial Drift Applied to the Annular Electron Beam

3.6.1 Creation of a Radial Drift

To create a radial $\boldsymbol{E}\times\boldsymbol{B}$ drift in the collector, either an E_θ or a B_θ field component is required. As discussed in section 3.2.2, the B_θ-type of MDC is excluded from this work, therefore, the key here is to create an E_θ field as shown in fig. 3.4a. In the previous work of this MDC type [68], a model with an infinite number of pipe-form electrodes was proposed. In this section, a more realistic two-stage collector design will be conceptually proposed.

The shape of the MDC design is based on the one of an axisymmetric SDC. A part of the SDC inner wall is a hollow cone as shown in the two-dimensional diagram of fig. 3.24. The conic inner wall is for the purpose, that the coil currents used to tune the magnetic field can be weak and the number of the tuning coils can be low (e.g. not more than 3). The coils tune the magnetic field in the SDC so, that the flux surfaces, where the electron beam is located, match the shape of the conic inner wall. The strongest tuning coil in fig. 3.24 overlays an additional field of 20 mT to the original collector magnetic field. Therefore, all of them can be built using normal-conducting wires. If the SDC would have everywhere an ordinary cylindrical inner wall (this case was also investigated in the simulations), it would require a higher number of much stronger coils to keep magnetic homogeneous, similar to the configuration in section 3.5.

The wall of the conic region is split by a helical gap into two segments as depicted in the perspective images at the upper part of fig. 3.24. The helical gap creates both of the electric field component E_θ for the drift and E_z for the deceleration. On the backside of the view, there is a straight gap between two ends of the helical gap. Beam electrons close to the helical gap drift radially outwards. The slow electrons stay a longer time in the drift field; therefore, they suffer from a larger drift distance and will be finally collected on the surface of the helical gap at the first stage. Fast electrons are less influenced by the drift and will be collected at the cylindrical top of the collector. However, the electrons at the azimuthal range of the straight gap drift inwards. This inward drift is unwanted and will be discussed later in the text.

As shown in fig. 3.24, the model of the electrodes has the base form of a thick metallic hollow cone. In the simulations, the thickness of the wall is chosen approximately between 5 mm and 10 mm depending on the range of D. The reason is that such a structure can facilitate the mesh generator in CST. Otherwise, a thin wall in the collector simulation would produce too many mesh cells for a structured volume mesh. In reality, the conic wall should be built with metal layers supported by frames, see the design in section 3.9.

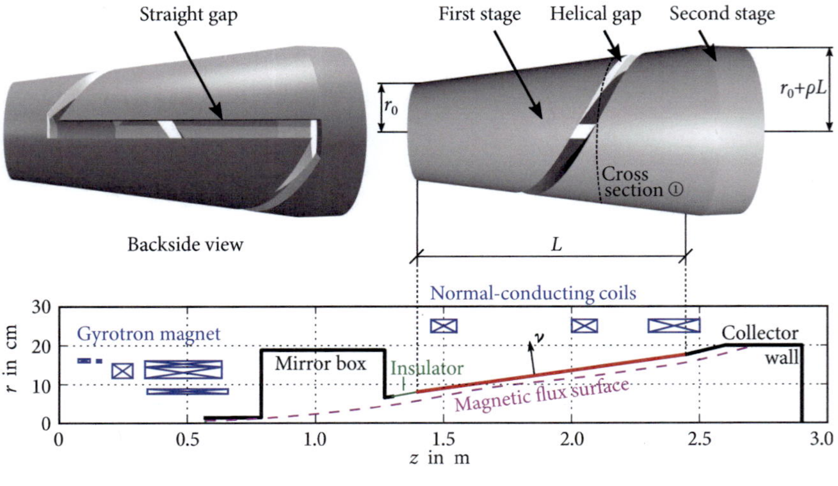

Fig. 3.24: Idea of an MDC with helical electrodes

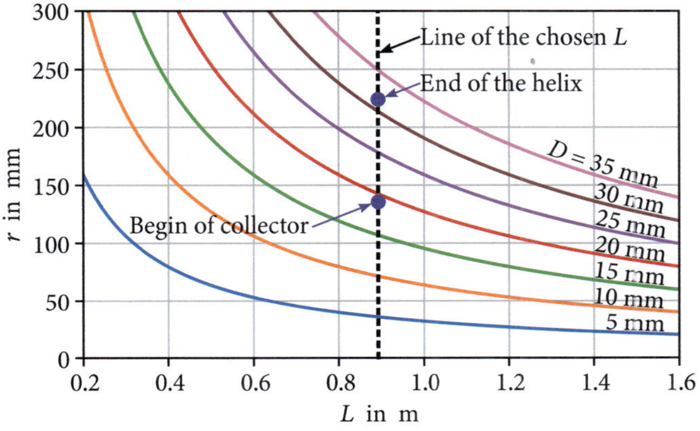

Fig. 3.25: The choice of design parameters based on eq. (3.34) [14]

3.6.2 Conceptual Collector Design and Basic Examinations

Choosing the Proper Design Parameters

The drift distance D of this proposal is similar to eq. (3.4). The difference is that the total magnetic field B, instead of its component $B_{\parallel}$, is considered in the calculation:

$$D = \frac{2\,v_0\,m_e}{B\,q}\tan\vartheta = \frac{2\,v_0\,m_e}{B\,q}\,\frac{\sqrt{1-\cos^2\vartheta}}{\cos\vartheta}\,. \tag{3.33}$$

According to appendix A.4, it is equivalent to

$$D = \frac{m_e\,v_0\,L}{q\,\psi}\,r = \frac{m_e\,v_0\,L\,\rho}{q\,\psi}\,l + \frac{m_e\,v_0\,L\,r_0}{q\,\psi}\,. \tag{3.34}$$

The newly involved variables (L, ρ and r_0) are marked in fig. 3.24, where $l \in [0, L]$ is the axial displacement of a point on the helix. Equation (3.34) means that the ***E***×***B*** drift distance increases linearly along the axis. Figure 3.25 is the diagram created from eq. (3.34) for the choice of design parameters. The height of the helix (L) is fixed at 0.9 m in order to have a comparable length as the reference SDC from fig. 1.11. The steepness (ρ) of the cone is chosen in the way, that the top radius of the helix (i.e. the maximal inner radius of the MDC) does not exceed 225 mm, which is the inner radius of the reference SDC. At the beginning of the collector, the drift is the weakest due to the high magnetic field. The minimum drift distance is set to be approximately 2 cm. Therefore, the distance between the electron beam and the conic wall is also kept around 2 cm.

In principle, a large drift distance is preferred. Two parameters determine the drift distance: the magnetic flux density B and the angle ϑ. Decreasing B or increasing ϑ will increase the drift distance D. However, the former will increase the collector radius; whereas the latter will decrease the torsion of the helix, thus the collector length will increase.

Examination of the Collector Efficiency

This MDC model is simulated using realistic electron beam parameters in CST, where the spent electron beam is imported from the interaction simulations with EURIDICE [48]. Results from both trajectory and particle-in-cell solvers are in very close agreement, where secondary electrons and space charges are taken into account in both simulation methods. The MDC model can theoretically achieve an η_{col} of 78 % (if the effect of secondary electrons is ignored) for $U_{d1} = 30$ and $U_{d2} = 42$ kV. The secondary electrons decrease the η_{col} only by 1 %. Details of the simulations and convergences are presented in [14].

Examination of the Cross Section to Understand the Behaviors of the Electron Beam

Figure 3.26 shows the cross section of the electron beam from the CST steady-state trajectory solver. The diagram demonstrates and verifies the principle of the operation. All electrons with $E_{kin} \geq 10$ keV are colored in red for a clear presentation. There is the helical gap between (c) and (d). The azimuthal location of this gap varies at different axial positions. The electrons at (f) are moving axially to the helical gap at their azimuthal angles. These electrons are not affected by the drift yet. Those electrons at (a) and (b) are the ones with high initial energies. They have been exposed in the **E**×**B** drift shortly. Although the drift has increased their radius, these electrons continue to move axially toward the second stage. The outer electrons pointed by (a) are the relatively slow ones.

They have been exposed in the drift for relatively longer time than the other ones at that angle. Hence, the radii of those electrons are larger than of the ones at (b). The electrons at position (c) start to be affected by the $\boldsymbol{E}\times\boldsymbol{B}$ drift. The ones at (d) are the originally slow electrons, which are pushed by the drift out of the main circular beam. They will be collected at the first stage.

The azimuthal position (g) is at the "backside" of the helical electrodes in fig. 3.24, where the drift direction is inverted. The angular range of this vertical gap is δ_1, whereas the sector of electron beam drifting inwards has an angular range of δ_2. The angle of the latter range is larger than the former. The electrons with low initial velocity at the angle δ_2 cannot overcome the potential barrier produced by the depression voltage, thus they will be reflected. These electrons cause a small back-stream beam current. A back-stream beam current of approximately 700 mA from the injected 45 A electron beam is observed in the simulations of this conceptual design. The discussion of this back-stream current will be given in section 3.6.3.

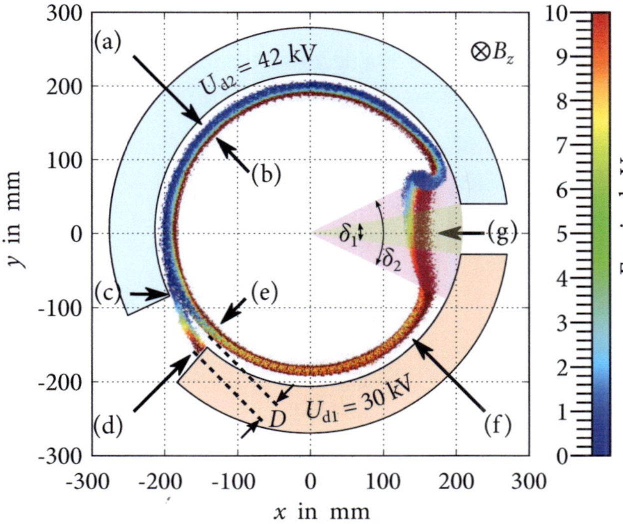

Fig. 3.26: Beam cross section (taken at position ① of fig. 3.24) in a steady-state trajectory simulation [14]

In the steady-state simulation, the same electron may pass through a cross section multiple times (even if it is not back streamed). For example, the slow electrons at position (e) also appear at position (d), as the drift is radial.

Examination of the Collector Performance Using Mono-Energetic Electron Beams

In addition to the simulations using the gyrotron spent electron beam with realistic energy spectra, the examinations with mono-energetic electron beams are also performed like in section 2.4.3. For this investigation, each simulation involves such an electron beam, where all electrons have the same initial energy. The thickness of the realistic electron beam is preserved. To compare fairly with fig. 2.9, the initial potential energies of injected electrons have to be considered.

Figure 3.27 shows the normalized current for the collection and reflection (back-stream). It proves that the ***E*×*B*** MDC performs the electron sorting very well. Compared to fig. 2.9, the energy sorting by the ***E*×*B*** drift is impressively sharp. That has two advantages:

- The sharp energy-sorting is a necessary condition for a high efficiency, as in this case, the electrons are collected at the most appropriate stages.
- Extending the number of collector stages for a sharp-sorting MDC can reward more efficiency than for an unsharp one.

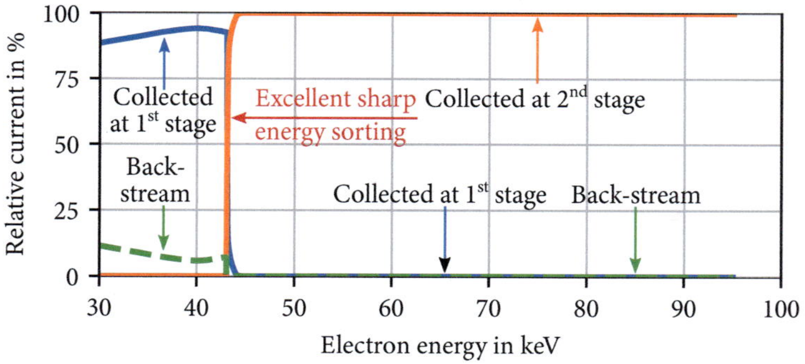

Fig. 3.27: Examination with mono-energetic electron beam (in contrast to fig. 2.9)

In contrast to fig. 2.9, this approach has no issue to sort the very high-energetic electrons. Hence, this concept is better than the one in chapter 2, if the issue of the back-stream current can be solved.

3.6.3 Assessment of the Critical Parameters

Achievable Collector Size and Mechanical Complexity

This MDC conceptual design is 10 % shorter than the reference SDC in fig. 1.11. However, its maximum outer radius (vacuum envelope not included) is approximately 10 % larger than that of the reference SDC. One advantage compared to any other MDC proposals is that, the required magnetic field in this proposal is very simple. One challenge to build such a collector is to create and cool the electrodes in helical strip form. This is feasible (see section 3.9) but need to be tested.

To have more than two stages, it is possible to stack the helical stages over each other [88]. Nevertheless, a two-stage collector could already have a sufficient efficiency for the first goal and should be considered for the construction and experimental verification, before extending it to more stages.

Expected Collector Efficiency

Simulations show that the two-stage collector has a collector efficiency of 78 % if secondary electrons are ignored, while $\eta_{\mathrm{col}} = 77\,\%$ is achieved in the simulation considering the effect of secondary electrons.

The $\boldsymbol{E}\times\boldsymbol{B}$ MDC shows a certain robustness against manufacturing tolerances and operation environments, because the moderate magnetic field (60 mT at the entrance, while 15 mT at the end) can stabilize the electron beam. In the following paragraphs, the sensitivities of the collector efficiency against

- beam current,
- beam misalignment, and
- external magnetic field

are checked by simulations like in chapter 2.

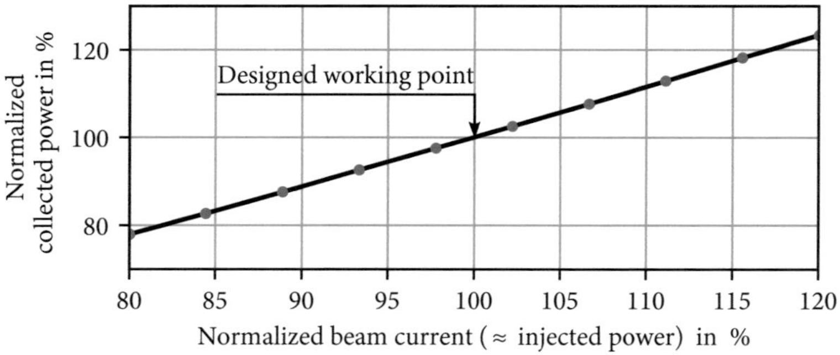

Fig. 3.28: Sensitivity of efficiency for ±20 % beam current [14]

Figure 3.28 shows the collected beam power over a certain range of beam current. The same kinetic energy spectrum is input in each simulation, whereas the beam current is scaled in a range of ±20 %. The curve is similar to the case of the axisymmetric MDC shown in fig. 2.12. It is estimated that the collected power varies with the beam current almost linearly, thus the collector efficiency is not notably influenced.

The check for the sensitivity of the beam misalignment is different from the check for the axisymmetric MDC in section 2.5.2. The difference is that, in this proposal, the transverse displacement of the electron beam affects the collector operation anisotropically (i.e. the effect of the beam misalignment depends on its direction). Figure 3.29a are the collector efficiencies presented in a polar coordinate. The "straight gap" is located at zero degree of the diagram, like in fig. 3.26. The simulations do not consider secondary electrons in order to be completed within reasonable time. Therefore, the reference simulation without misalignment has a slightly higher efficiency than the simulation with secondary electrons presented before. An electron beam displacement of 0.5 mm in the cavity influences the collector efficiency by ±1 %. When the beam misalignment is increased to 0.8 mm, in the worst case, the collector efficiency is reduced by 4 %, which is much better than the axisymmetric MDC (presented in fig. 2.11).

The tolerance of an external perturbation magnetic field is also checked in the simulations ignoring secondary electrons. The perturbation field includes the stray field of tokamak coils and the magnetic field from neighboring gyrotrons.

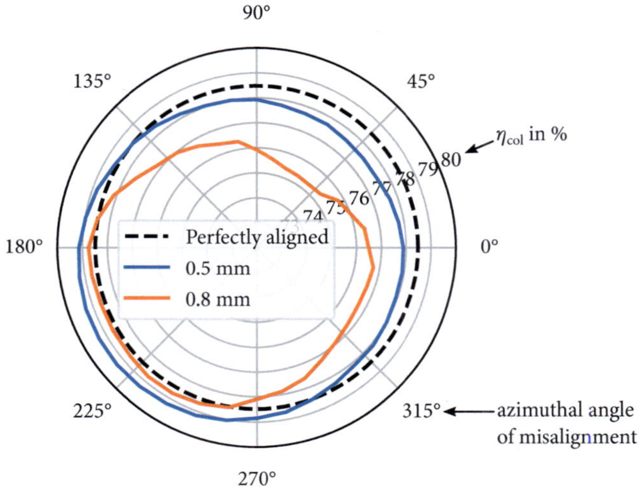

(a) Misaligned beam in cavity

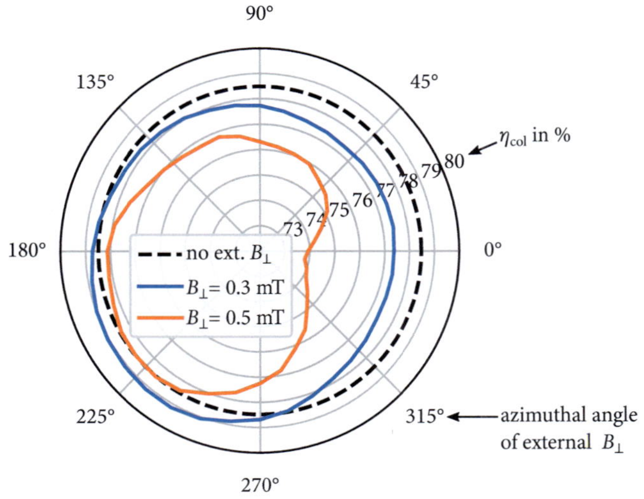

(b) Perturbation of $B_\perp$

Fig. 3.29: Sensitivities of efficiency for anisotropic misalignment and perturbation $B_\perp$ [14]

Since the axial component of the external field can be theoretically compensated by the collector coils, only the influence of the planar transverse magnetic field (magnetic flux density) $B_\perp$ is investigated. Depending on the direction of the applied $B_\perp$, an anisotropic diagram is depicted in fig. 3.29b. In the worst case with $B_\perp = 0.5\,\mathrm{mT}$, the collector efficiency is reduced by 5 %. In the best case (if the collector is mounted with the optimal azimuthal direction), the $B_\perp = 0.5\,\mathrm{mT}$ will have no influence on the collector efficiency. Compared to the same study for the axisymmetric MDC in fig. 2.10, which has 8 % efficiency reduction in any direction, this $\boldsymbol{E}\times\boldsymbol{B}$ concept is advantageous.

As expected, the collector efficiency of this proposal has certain tolerances to the manufacturing and operation environment. However, only the collector efficiency was checked in this investigation. Impact of the back-stream beam current and the changes of the power load density are also important and will be considered in future.

Unwanted Back-Stream Beam Current

The major physical uncertainty of this MDC proposal is the impact of the back-stream current. In this particular design, 1.5 % $\approx$ 700 mA current of the input spent electron beam is back-streamed. The diagnostic simulation in fig. 3.27 shows that the reflected electrons are a minor part ($\approx$ 10 %) of the slow electrons which should be collected at the first stage. No electron, which should be collected at the second stage is reflected. According to the energy spectrum of the spent beam in fig. 1.9, the proportion of electrons which have relatively low energy (30–42 keV) is very small. In other words, the reflected current is just *a minor part of another minor part* of the primary electron beam.

A theoretical study on the impact of back-stream beam current is based on the full-gyrotron simulation with a regular SDC of the 170 GHz 1 MW gyrotron using ARIADNE. (ARIADNE is not yet capable of modeling this MDC shape at the time of writing, therefore, an axisymmetric SDC, which produces the same level of back-stream current, is applied in this study.) It is found that the back-streamed electrons will re-participate in the interaction and will be eventually accelerated by the RF field. They have less chance to move to the cathode due to the low potential there. More probably, these electrons gain energy from the interaction, then they re-enter the collector at other azimuthal angles and will be collected appropriately. Following are the details of this investigation.

In the full-gyrotron simulation, the depression voltage is intentionally set to an exaggerated value. This causes a back-streaming current of 1.2 A from the 45 A primary beam current. The evolution of the RF output power towards a steady state is shown in fig. 3.30. In this simulation setup, the first ten (an arbitrary number as far as the first steady state is reached) iterations do not consider the interaction with RF, in order to obtain an intermediate steady state of the beam optics. For this intermediate steady state, a conservative depression voltage is applied, so that no electron should be reflected from the collector. Starting from the 11th iteration cycle, the interaction with RF is turned on. After another 13 iterations, a new intermediate steady state is reached regarding the self-consistent interaction simulation. Afterwards, the depression voltage is switched to an exaggerated value, which forces the collector to back-stream a beam current of 1.2 A. After a certain number of iterations, the RF power reaches another steady state. The new power-level is marked by the red dashed line in fig. 3.30. It is found that even with such a high back-stream current (70 % higher than the value predicted for the MDC proposal), the reduction of the output power is insignificant, and the reflected electrons are re-collected in the collector.

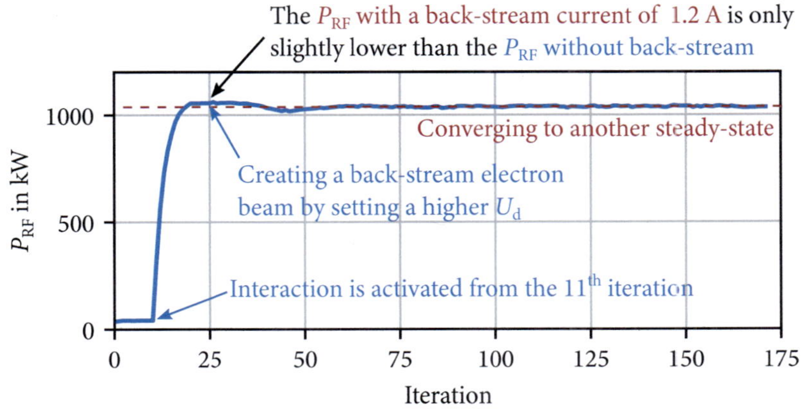

Fig. 3.30: Convergence of the RF power in the ARIADNE full-gyrotron simulation with 1.2 A of 45 A back-stream current

Experimentally, there are two particular verifications presented in the literature:

- A back-stream current of 10 mA is measured in the experiment of the 110 GHz gyrotron [71]. However, 640 mA would be reflected from the collector according to simulation. This could mean that the rest of the back-stream electrons reenter the collector with a higher energy and are collected eventually.
- For the 140 GHz gyrotron [72] with $I_\mathrm{b} = 27\,\mathrm{A}$ and $U_\mathrm{b} = 80\,\mathrm{kV}$, the measured RF output power does not change, even when the measured reflection of electrons reaches 3 A.

From both experiments one can expect that a back-stream current of the simulated quantity does not affect the gyrotron operation. Nevertheless, to avoid any unexpected effect of the back-stream current, solutions to suppress this current will be found. There are two options to reduce the back-stream current without changing the collector basic geometry too much:

- Decreasing the width of the straight gap can reduce the back-stream current. The gap should be at least so wide that there is no breakdown of the electric field.
- As shown by fig. 3.27, the relatively faster electrons of those slow electrons (e.g. the ones with $E_\mathrm{kin} = 40\,\mathrm{keV}$ in fig. 3.27), which should go to the first stage, are less probably to be reflected. Therefore, decreasing the depression voltage at the first stage can also slightly reduce the reflection. However, the price for that is a reduction of the collector efficiency and an increment of the maximal power density at the first stage.

Instead of tuning the design parameters, there are advanced proposals which suppress the back-stream current by their operation principles. Such proposals will be introduced in section 3.7.

Maximal Expected Power Load Density on the Critical Parts

The power density information is given by CST simulations and can only be processed inside this commercial software. That limits the flexibility of the result evaluation, especially in the current CST implementation of the power load density function. Hence, the presented results in the following paragraphs are preliminary estimations.

The power load density study is based on the 170 GHz 1 MW gyrotron as for the other proposals presented in previous sections. At the first stage, where the electrode is a helical surface, the power is concentrated on the inner side of the helix, as shown in fig. 3.31. The edge lengths of the triangle mesh surfaces (of the unstructured tetrahedral mesh cells) are approximately 1 cm. Most cells have power density below 200 W/cm^2, while the hot spots are below 300 W/cm^2. On the second stage shown in fig. 3.32, the load is higher. Hot spots are approximately 700 W/cm^2, which exceed the limit. However, the 700 W/cm^2 here is the static load, while the 500 W/cm^2 limitation is for the temporally averaged load with a sweeping of a 2 kW/cm^2 electron beam. Hence, how does the 700 W/cm^2 static load affect the collector fatigue has to be further investigated. A thermal analysis [23] shows that a power density of 900 W/cm^2 means a temperature of 282°C if the power can be removed within 0.5s.

There are a few possibilities to reduce the 700 W/cm^2 power load density at the last stage:

- Increase the collector inner radius by 40 % to be approximately 320 mm.
- Upgrade the MDC to have a higher number of stages.
- A local beam sweeping similar to the SDC can be additionally designed. Since electrons experience more deceleration at the last stage of the MDC than in an SDC, the requirement of the sweeping system in this case can even be lower than that in the reference SDC.

In case of a future DEMO 2 MW gyrotron, the beam power is doubled. At the first stage the power load may still be under the limitation. At the second stage, the combination with a beam sweeping is necessary.

Above are preliminary studies. The values of load vary with the numerical settings and the results presented in this study are quite limited by the existing software implementations. Further analysis and thermal simulations should be performed to have a better understanding of the issues and to propose better solutions.

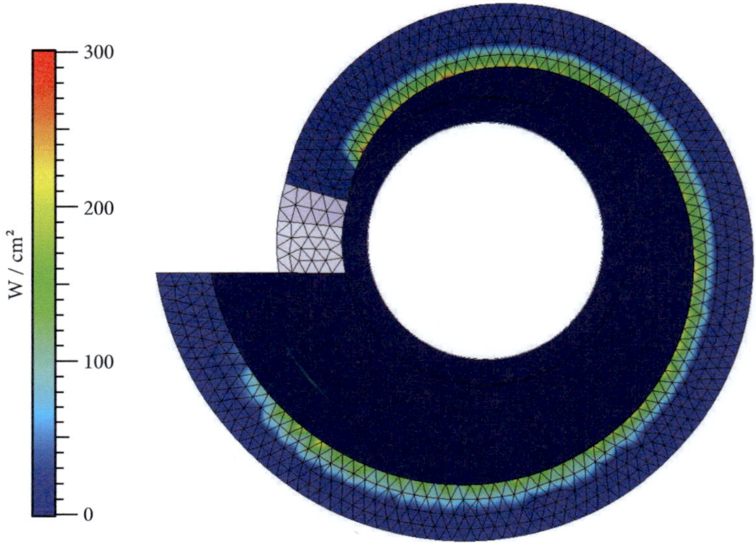

Fig. 3.31: Power load density at the first stage

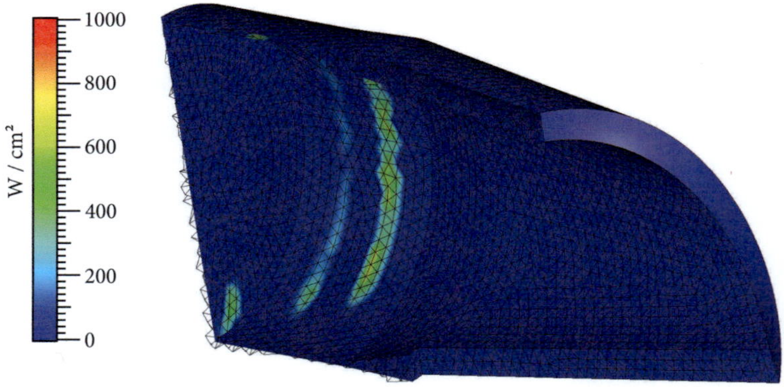

Fig. 3.32: Power load density at the second stage

Summary

This MDC proposal requires the simplest collector coils among all existing gyrotron MDC proposals. It also leads to a most compact conceptual design of a two-stage collector with a theoretically sufficient collector efficiency, which is less sensitive to the beam misalignment and the perturbation magnetic field, compared to a classical one. The power load study is limited by the implementation of the used commercial software. This power load might be controllable for the future DEMO gyrotron, if a beam sweeping would be applied at the last collector stage.

The conceptual weakness of this proposal is the existence of a back-stream from the primary electron beam current. This weakness will be remedied in the upgraded proposals of section 3.7.

3.7 Upgrades for Proposal IV

3.7.1 Upgraded Design Proposals to Suppress the Back-Stream Current of Proposal IV

There are multiple reasons which can cause a part of the beam electron to be back-streamed. The substantial back-streaming of beam current is caused by the flip of drift direction at the azimuthal angle, where the straight gap (see fig. 3.26) is located. Four upgraded design proposals for significant reductions of the back-stream are proposed in this section. All of them are aimed to reduce this type of reflection.

Upgraded Proposal 1: Introducing Another Azimuthal Drift

The first proposal for the minimization of the back-stream is to prevent slow electrons from staying at the azimuthal angle of the straight gap. These slow electrons can be carried by an additional drift to other azimuthal positions. Furthermore, there should be no other electron filling back to the gap.

Inserting another electrode can perfectly create such a drift. The principle is shown in fig. 3.33. The figure is a perspective bottom view of the cross section in the middle of the upgraded MDC proposal. The outer structures are the original electrodes. The inner one, which has the electric potential of the second stage, creates a radial electric field only on the right half side of this figure. In order to facilitate the mesh generation, the inserted electrode is schematically bulky just like the outer electrodes, whereas a real implementation would only need its electric potential on the boundary. The orange ring between the outer and inner electrodes marks the cross section of the annular electron beam.

Those electrons on the azimuthal angle of the straight gap experience a clockwise drift in the view of fig. 3.33. The drift rotates these electrons along the equipotential surfaces to another azimuthal position. There, the ***E***×***B*** sorting will take place after an axial displacement at a higher position. The electrons, which were originally reflected in the proposal of section 3.6 will be appropriately collected (they follow that magenta trajectory). As a side effect of this upgrade, the newly introduced drift also pushes the electrons on the right half side of fig. 3.33 faster to the helical gap, which anyhow does not affect the operation principle of the original MDC.

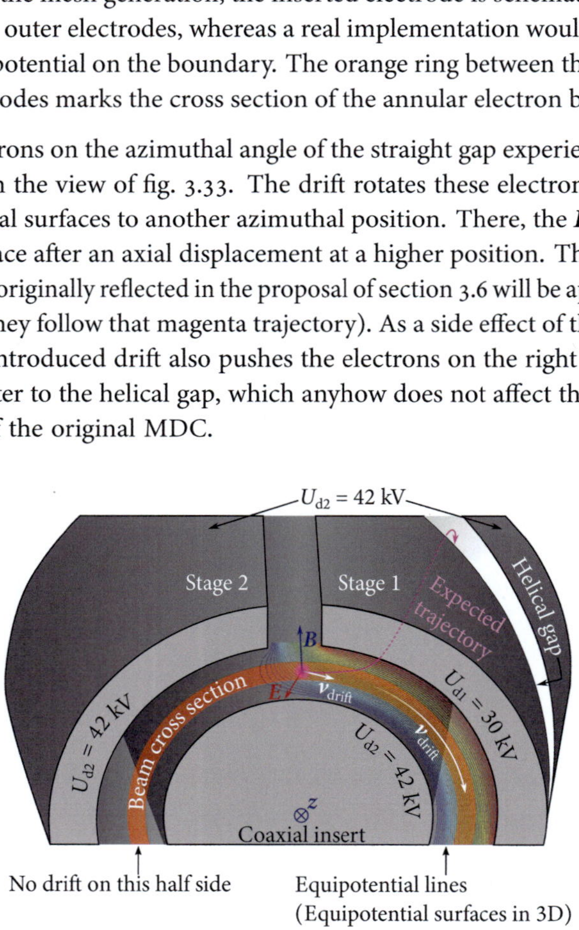

Fig. 3.33: Azimuthal drift produced by a coaxially inserted electrode [15]

The left half side of the electron beam in fig. 3.33 has already been sorted by the helical electrodes. This part of the remained electron beam only has electrons with high initial energies. In the absence of space charges, there is no azimuthal drift for those electrons. They will stay in that azimuthal angle until they reach the second stage. Even if the space charges produce such an azimuthal drift, that this part of the electron beam is rotated to the azimuthal angle of the straight gap, those electrons will still not be reflected due to their high energy.

The upgraded design proposal seems to be a reasonable solution for minimization of the back-stream current. However, the simulation with CST still predicts 1.1 % (500 mA of 45 A) reflected beam current (including secondary electrons), which is nevertheless better than in the original proposal, where the ratio was 1.5 %. Two reasons are found for the explanation of the unwanted back-stream current:

- The coaxial insert increases the absolute value of depression voltage at the beam radius. The new depression voltage reflects the slowest part of the beam electrons before they can be sorted by the $\boldsymbol{E}\times\boldsymbol{B}$ drift. Hence, the depression voltages have to be optimized differently for this upgraded proposal.
- The added azimuthal drift is not effective enough. It is observed in the simulation, that some slow electrons still stay in the azimuthal range of the straight gap. Further optimization of electrode geometry and voltages might also improve this situation.

Since the depression voltages, geometry and magnetic field have to be tuned carefully in order to further reduce the back-stream current, the resulted $\boldsymbol{E}\times\boldsymbol{B}$ MDC design will be less flexible on the choice of the design parameters. To simplify the geometry, the design with inverted inner and outer structures presented in appendix B.3 is another option for the implementation of this proposal.

Upgraded Proposal 2: Increasing the Angular Range of the Helical Gap

In the original MDC design, the angular range of the helix was less than 2π, where the straight gap causing the reflection also takes a tiny range of the azimuthal angle. The following proposal hides the straight gap behind a piece of the helical electrodes. Any electron arrives the straight gap should have already been sorted by that piece of the helical electrodes.

Figure 3.34 shows the principle of this proposal. Here, the helix twists a bit over 360°. The original straight gap is split into a short segment and a long one. The short segment has a width of d_1, which is wider than the width d_2 of the long segment. Since the azimuthal electric field is weaker with a wider gap, the drift velocity is also slower at the short segment according to eq. (1.15).

At the azimuthal position marked by the green arrows, electrons are sorted at the entrance of the collector. The slow electrons are split out and collected at the first stage immediately after they enter the MDC. Only fast electrons are in the rest of the beam at that azimuthal position. Although there will be an inward drift until the end of the straight gap, the fast electrons will not be reflected.

At the angle marked by the white arrows, the electrons entering the collector pass through the short but wide gap. The large width results in a slow inward drift, as discussed above. The short length means a short exposure time for the electrons in this drift field. Moreover, the magnetic field at the entrance is four times higher than at the end of the collector, therefore, the drift velocity at the entrance is four times weaker. Due to the low electric field, short exposure time and high magnetic field, electrons at this angle only experience a very weak inward drift, such that they will not drift too far from the collector wall. After the short gap, electrons continue their movement until the helical gap. As the electrons reach the helical gap, they are decelerated by the longitudinal electric field, so that the exposure time in the ***E***×***B*** drift is long, especially for the slow electrons which may be completely decelerated. In addition, the magnetic field here is weaker. The outward drift by the helical electrodes, which performs the energy sorting, is much stronger than the previous inwards drift at the wide gap. Hence, the electrons at this angle can also be appropriately sorted and collected.

Depending on the setup, this upgraded design still have 150 ± 80 mA back-stream current from the total 45 A injected electron beam. The improvement to the original 700 mA is significant. Two kinds of electrons might be reflected in this particular simulation model:

- The first kind is the extremely slow electrons at the azimuthal angles marked between the green and white arrows. They cannot overcome the electric potential and are totally decelerated. During the deceleration, they drift inwards and are reflected.
- The part of the electron beam pointed by (g) in fig. 3.26 experiences the inward drift. However, the affected electrons are not only the ones exactly at the azimuthal range δ_1. The effective azimuthal range, where electrons drift away from the collector wall, has the angle of δ_2 in fig. 3.26. The value of δ_2 increases with the length of collector. If the collector is long, the range δ_2 may include those electrons at the angle marked by the white arrow in fig. 3.34. The slower electrons are more affected, such that they cannot optimally drift to the first stage and will contribute to the back-stream current, too.

Two solutions can suppress the second kind of back-streaming. One is to further twist the helix; however, that would increase the collector size. The other is to reduce the collector length without decreasing the outward drift distance (will be presented in section 3.7.2), in order to reduce the exposure time of electrons in the unwanted drift field.

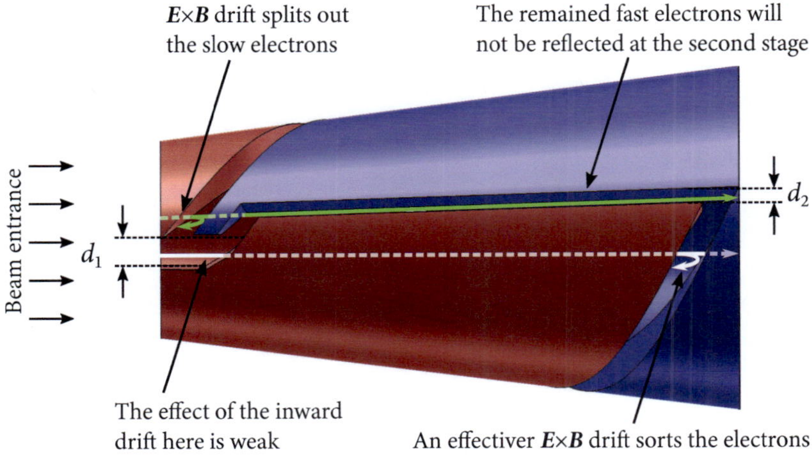

Fig. 3.34: The helical structure covers more than 2π azimuthal range [15]

Upgraded Proposal 3: Capturing the Back-Stream Electrons Using a Disk

Since the back-streamed electrons drift inwards, they have notable smaller radii than the input electron beam. These electrons can be captured using a disk, which is placed within the electron beam. This disk should have the same potential as the first stage. The topology determines that this disk can only be hung from the top of the collector. The originally back-streamed electrons are expected to have the yellow trajectory in fig. 3.35 Due to the 60 mA magnetic field at the collector entrance, the beam electrons have small Larmor radii, i.e. the annular beam is thin and well confined. The disk is supposed to be placed close to the beam at a distance of 1–2 cm.

The disk reduces the back-stream current from originally 700 mA to 200 mA. These back-streamed electrons are the ones having similar radius as the input electron beam, so that they can backwards pass through the gap between the disk and the collector wall. For the 170 GHz 1 MW gyrotron there can be 2.4 kW beam power (including secondary electrons) absorbed by the disk. This power should be dissipated by water, which can be fed through the hanging structure. The power absorbed at the disk can be reduced, if the disk is combined with the former proposal.

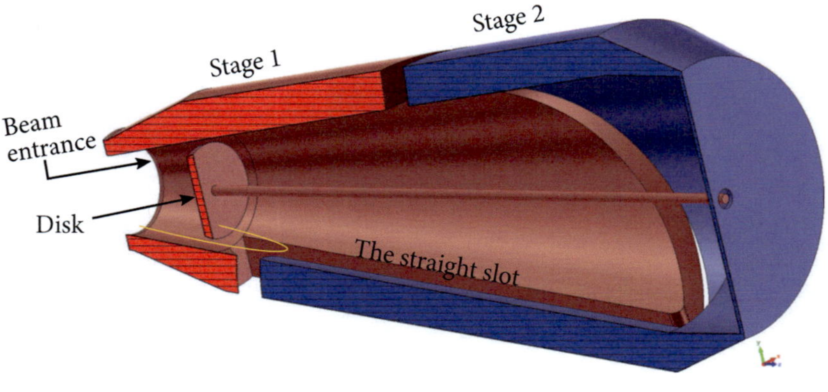

Fig. 3.35: Disk at the entrance catching the reflections [15]

Upgraded Proposal 4: A Combined Solution

The combination of upgraded proposal 2 and 3 is conceptually investigated in the following paragraphs. Although the functionalities of the single proposals are not orthogonal, the combination integrates the advantages of both upgraded proposals. In particular:

- The combined proposal can further reduce the back-stream current for a factor of ten.
- Compared to the original upgrade 3, the cooling requirement of the disk is relaxed.

The geometry of the collector entrance is shown in fig. 3.36. Simulations using the realistic spent electron beam of the 170 GHz 1 MW gyrotron predict in the worst case only 20–30 mA back-streamed secondary electrons (In the simulations without secondary electrons there is no back-streamed current. Even in some cases involving secondary electrons, there is an $I_{\text{ref}} < 1$ mA without any reduction of the collector efficiency.) An ideal extrapolation to the 2 MW DEMO gyrotron means to have a back-stream current up to 60 mA. In the worst case, 100 mA (corresponding to 1 kW) of the electron beam is collected at the disk, which is roughly half of the current predicted in upgraded proposal 3.

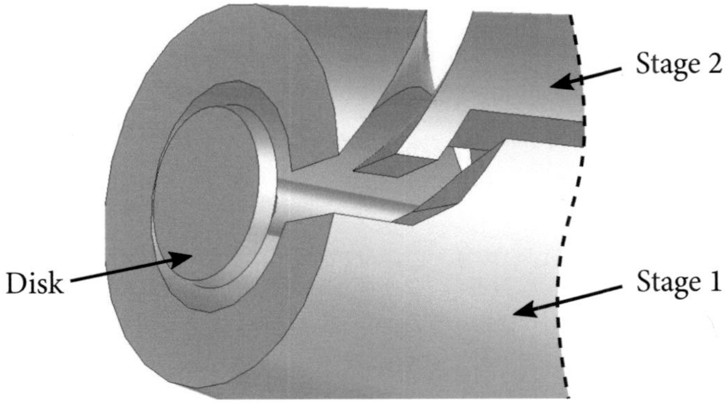

Fig. 3.36: Combination of figs. 3.34 and 3.35 [15]

To check how well the MDC reacts to the electrons of different initial energy, a set of simulations considering mono-energetic electron beam are performed for this upgraded MDC proposal like in sections 2.4.3 and 3.6.2. Figure 3.37 shows the beam current distribution at each stage. The simulations here do not involve secondary electrons mainly due to the cost of simulation time. For the realistic spent electron beam having an energy range from 38 keV to 120 keV, no electron is back-streamed according to fig. 3.37. As expected, the disk only collects the electrons, which have initial energy below the depression voltage of the second stage (i.e. the blue curve goes to zero when the electron energy is high).

Figure 3.37 also shows that the energy sorting by the $\boldsymbol{E}\times\boldsymbol{B}$ drift in this MDC proposal is not as sharp as the one shown in fig. 3.27. The reason could be an imperfect optimization. The torsion and the length of the helix in this upgraded proposal are greater than in that one (in order to have a conservative model to ensure that the principle works). However, the other parameters are unchanged. Therefore, the drift distance here is longer, such that the radius of the conic part does not match the new drift distance. Hence, not only the slow electrons, but also some fast ones may drift to the helical gap and collected at the first stage. A new optimization of the design parameters should improve the situation.

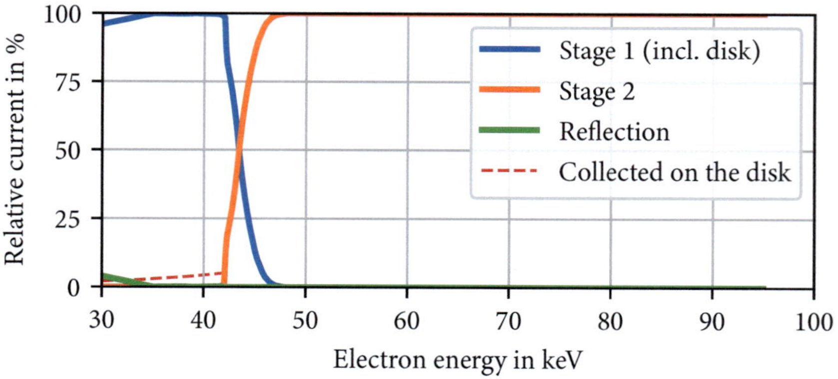

Fig. 3.37: Examination with mono-energetic electron beam

3.7.2 The Final Upgrade for a Reduction of the Collector Size

Since the back-stream current is suppressed using the last upgraded proposal, there is now no issue to azimuthally repeat the helical gaps and the straight gaps. For n repetitions of the gaps (the case of $n = 3$ is given in fig. 3.38), the $\boldsymbol{E}\times\boldsymbol{B}$ drift region of the new collector is theoretically only $1/n$ long of the original one. This means that the last design in section 3.7.1 can be shrunk to a smaller one like in fig. 3.38, which has approximately $1/3$ length of the original one. The maximal inner radius of the MDC is reduced with the length, too. The scaling factor of the maximal inner radius depends additionally on the beam radius at the entrance, because the beam radius at each axial position is not supposed to be scaled.

A beneficial side effect of reducing the MDC size is that, the collector coil system can be remarkably simplified from two aspects. On one hand, since the collector is shorter, less coils are required to tune the magnetic field for the shorter collector length. The total collector coils can be lighter. On the other hand, the field of the gyrotron main SC-magnet decays in the order of z^{-3}, as shown in section 3.5, rather than quadratically. Therefore, the magnetic field lines in the collector have a curvature. It is easier to collimate the field lines to be aligned with the conic wall, if the collector is short.

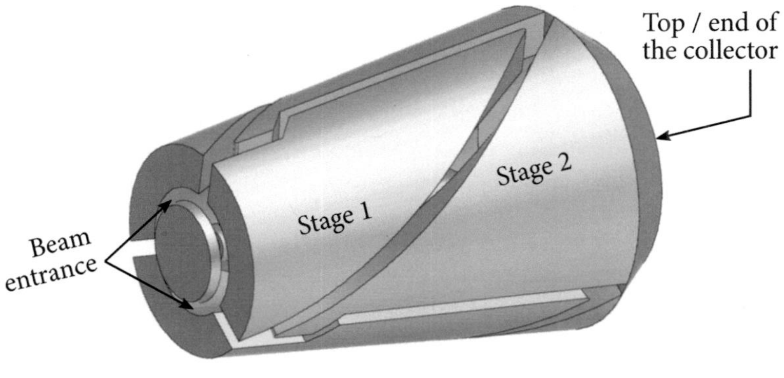

Fig. 3.38: Azimuthal repetition of helices [15]

This final upgrade has another unique advantage. Since the collector is shorter than before, the time an electron requires to move from the entrance of the collector until the end of a helix is reduced, i.e. electrons spend less time at the azimuthal angles where the **E**×**B** drift is inward. The angular range δ_2 in fig. 3.26 is much smaller than in the full-length designs and δ_2 can be close to δ_1. Hence, the slow electrons near the (long segment of the) straight gaps are better handled in this upgraded proposal.

The conceptual design in fig. 3.38 is simulated using the spent electron beam of the 170 GHz 1 MW gyrotron. The number of helices does not influence the collector efficiency. The reflected current of primary electrons is 22.5 mA assuming an input current of 45 A. Including the secondary electrons, there will be a total current of 41.4 mA back-streamed. The back-stream current of secondary electrons can be further reduced by tuning the shape of the top cylinder. A current of 37 mA $\approx$ 400 W from the primary electron beam is collected at the coaxial disk.

3.8 Comparison of the Design Proposals in this Chapter

All design proposals presented in this chapter are compared in table 3.1. The calculated maximal power load densities are not compared in that table, since the evaluation of the power density is not well implemented in the available software. (Particularly, the exact numerical value of the power density depends strongly on the interpolation of the meshed geometry and of the spatial beam-power distribution.) Among all design proposals based on the **E**×**B** drift:

- Proposal I (section 3.3) is ideal for an energy-sorting using the **E**×**B** drift. But it is not the best option for the gyrotron MDC experiment, because it is huge (for the reference 170 GHz 1 MW gyrotron as well as for the target DEMO gyrotron) and requires special coils.
- Proposal II (section 3.4) will also result in a very large collector size for the reference and the targeted DEMO gyrotron.

Table 3.1: Comparison of the MDC design proposals based on $\boldsymbol{E}\times\boldsymbol{B}$ drift

Proposal		I	II	III	IV	Final upgrade
MDC size	Relative length*	> 130 %	Too huge for the 170 GHz 1 MW gyrotron	130 %	90 %	1/N
	Maximal width of the inner walls	1.2 m		0.63 m	0.6 m	< 0.6 m
Complexity	Coils	at least 9 long coils		4 moderate solenoids	3 weak solenoids	≤ 3 weak solenoids
	Electrodes	Straight pads		Planar strips	Helical strips in 3D	
Extending to more stages		Straightforward		Straightforward	Possible [88]	
η_{col} of two stages**		80 %		74 %	78 %	
Back-stream current*** of an input $I_b = 45$ A		No		Negligible	700 mA	40 mA

* Compared to the reference SDC   ** Secondary electrons ignored   *** Secondary electrons included

- Proposal III (section 3.5) is better than proposal II. However, it is also not the best option, because the baffles limit the achievable efficiency and cause hotspots of the power load.
- Proposal IV is reasonable for implementation of a DEMO gyrotron MDC. This proposal has even attracted the interest of other researchers, such that the theoretical operability has also been confirmed in [106, 107].
- The final upgrade of proposal IV suppresses the back-stream current and shrinks the collector size. This will be considered for the mechanical design of the world-first gyrotron MDC.

3.9 Proposal for the Mechanical Design of a Two-Stage Depressed Collector

The comparison of all MDC concepts based on $\boldsymbol{E}\times\boldsymbol{B}$ drift in section 3.8 leads to the conclusion, that the first proof-of-concept MDC experiment should consider the final upgrade of design proposal IV (the last column). A collector consisting of two stages is preferred for the first try, even if the $\boldsymbol{E}\times\boldsymbol{B}$ concept is more likely to show its real advantages with numerous stages. This section shows the mechanical design of that best approach. Since the exact parameters for the DEMO gyrotron, especially the profile of the SC-magnet is not finalized at the time of this work, the mechanical design presented in this section is based on the 170 GHz 1 MW gyrotron. Short-pulse (milliseconds) operations are foreseen for the initial proof-of-concept experiments. The beam sweeping at the second stage is not included in a short-pulse operation. Nevertheless, in order to operate with a high duty cycle, the water cooling is possible and is also optionally considered at both stages to actively dissipate the heat.

Figure 3.39 gives an overview of the mechanical design. The first stage of the MDC is insulated in the same way as the reference SDC, a circuit of the latter is shown in fig. 1.2. The mirror box is grounded and insulated from the body potential, while the first stage and the vacuum envelope (c) of the MDC are electrically connected to the top of the mirror box (a). Component (d) is the helical surface at one of the three electrodes for the first stage.

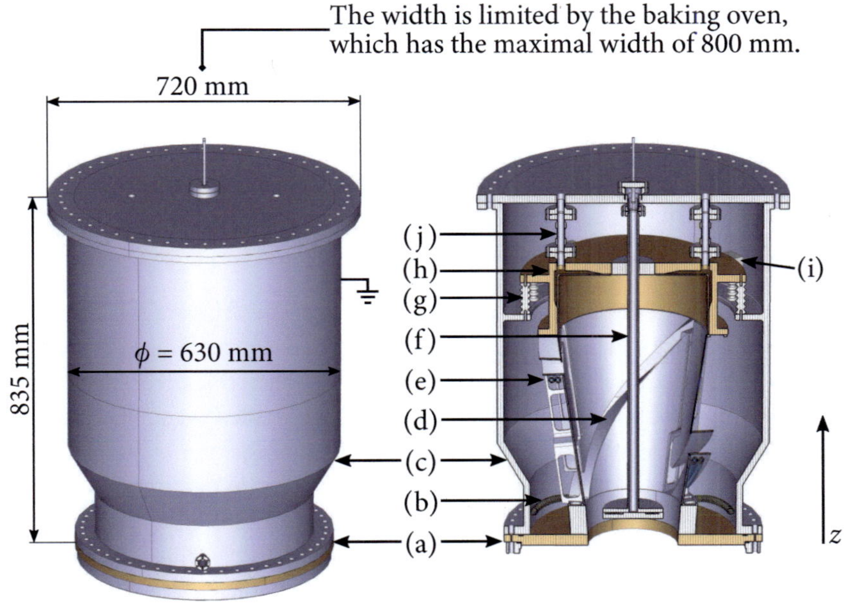

Fig. 3.39: Mechanical design proposal of a two-stage depressed collector. Components (a)–(f) are grounded and insulated from the gyrotron body (cavity) potential (ideally +30 kV). The second stage (h) at −12 kV is insulated by the insulators (g) and (i).

Fig. 3.40: A module of the first stage

The cooling water for the first stage is fed from the inlet (the "navel" in the left picture) through the vacuum envelope, and then guided by pipes (b) to the cooling pipes (e) beneath the helical surface. Theoretically, electrons can be collected directly at the cooling pipes; however, there are tiny gaps between two adjacent pipes, where electrons can pass through. Therefore, the helical surface (d), which can be constructed piecewise, should cover the pipes to capture all electrons going to the first stage. The first stage consists of three identical modules (one of them is shown in fig. 3.40).

The lower part of the second stage consists of bent metal plates, which create with the first stage together an E_θ. The upper part of the second stage (h) consists of two components: an outer metallic part, which has a thick cross section in this cut view, and a thin inner cylinder. They are welded together. Between them there is a chamber for the cooling water. Cooling water is pumped into the chamber via one of the insulated pipes (j) and flows out via another pipe at the other side. Between both ends of (j) there is a voltage of up to 12 kV, while (j) is chosen to be capable of insulating a voltage of 20 kV. The voltage of the second stage can be fed from the top or side of the vacuum envelope (not visible in the figure). The ceramic insulators (g) are fixed at the second stage. During the assembling (see fig. 3.41), the entire second stage including the insulators are plugged into the vacuum envelope, which lifts the weight of the second stage. A transverse movement of the second stage inside the vacuum envelope would break the insulators (g). To limit this movement, multiple ceramic blocks (i) should be placed at multiple azimuthal angles. They fix the transverse position of the second stage.

A hollow coaxial insert (f) is hung from the top of the vacuum envelop. It is also on the ground potential. Cooling water for the disk can be fed through the hanging structure.

The final collector will weigh totally in the order of 300 kg. It is about 30 % shorter than the reference SDC. The outer radius of the vacuum envelope is 315 mm, however, the maximal radius of the envelope is determined by its top flange, which is in this design 360 mm. The collector size does not exceed the limitation of the oven available at KIT.

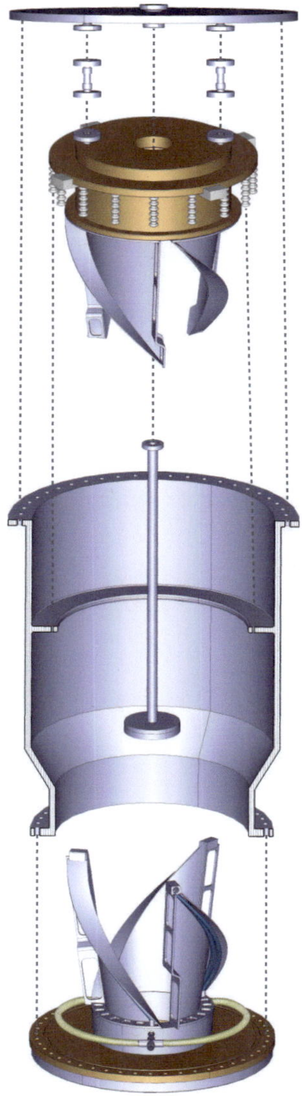

Fig. 3.41: Assembling the components of the mechanical design proposal

4 Conclusion and Outlook

The Multistage Depressed Collector (MDC) is the key to achieve the gyrotron efficiency required by the DEMO and future fusion power plant. Although MDCs were successfully applied in other kinds of vacuum tubes (especially in TWTs) since decades, there is yet no MDC operating in gyrotrons. One remarkable reason is that there is a moderate magnetic field in the gyrotron collector region, which strongly confines the spent electron beam. This magnetic field prevents the electrons from spreading by the repulsive force between the negative charges. For this reason, the possibility to use an electrostatic lens for the sorting of the electrons is excluded. Another remarkable fact is that, the annular small-orbit spent electron beam encloses a large magnetic flux. Reducing or shielding the magnetic field until the electric field is dominating will result in an extremely large collector size. Therefore, new possibilities to apply the multistage depressed collection should be invented and systematically investigated.

Conclusion of the Systematic Investigations

There are two fundamentally different kinds of gyrotron MDC concepts. The first kind is axisymmetric. The other one relies on the $\boldsymbol{E}\times\boldsymbol{B}$ drift of beam electrons. In order to create a useful $\boldsymbol{E}\times\boldsymbol{B}$ drift, the axisymmetry has to be broken. Both kinds of MDC concepts are systematically investigated in this work. A wide variant of design possibilities is proposed for the conversion of the physical principles into collector conceptual designs.

The most researches on gyrotron MDCs prior to this work belong to the axisymmetric concept. The biggest advantage is obviously the symmetry, so that the description and analysis of the collector system are two-dimensional. Thus, the construction is supposed to be also simpler than an antisymmetric design. A two-dimensional description of the geometry, electron trajectories and fields is especially appropriate, as the computational resource and technologies for the mechanical construction decades ago were not such flexible and multi-variant as today. However, a theoretical analysis of the axisymmetric concept immediately shows its weakness that is the conservation of the canonical angular momentum (Busch's theorem) does not allow the separation of electron orbits.

To explore the limitations of axisymmetric concepts, a new axisymmetric design proposal based on a different principle is presented. That proposal is more suitable for high-frequency gyrotrons than the state-of-the-art axisymmetric ones. In the state-of-the-art proposals e.g. the axisymmetric collector proposed in [10], the magnetic field should be reduced until the electric field dominates the control of trajectories. This is not practical for high-frequency fusion gyrotrons due to the large collector size caused by the gently decreasing of magnetic field. The new proposal presented in chapter 2 is based on the phenomenon, that a strongly non-adiabatic transition of magnetic field sorts (modulates) the local phases of the electron-cyclotron motions. A conceptual design based on this physical principle requires significantly less space than the up-scaled state-of-the-art axisymmetric ones. Moreover, the magnetic field can even spread the electron beam statically over a large area at the collector wall, such that the sweeping of the electron beam like in an SDC is not necessary for the reference 170 GHz 1 MW gyrotron. Extrapolating the design proposal to the DEMO gyrotron which will output an RF power of 2 MW, the collector radius should be doubled (larger than 0.5 m) in order to scale the maximal impacting power density below 500 W/cm^2.

It is shown theoretically that this kind of sorting mechanism cannot sort electron energy precisely, although it requires a precise magnetic field. Simulations predict that a two-stage axisymmetric collector can almost achieve the goal of 60 % gyrotron overall efficiency, but only if there would be no influence of secondary electrons. Simulations taking into account the secondary electrons show a much lower efficiency. Upgrading the design to more than two stages may theoretically increase the achievable efficiency, however, that will be very challenging. The root for the large influence of secondary electrons and the limited upgrade possibility is that there is no real separation of electrons in an axisymmetric system. Hence, this concept may not be the best option for an MDC targeting the DEMO gyrotron.

The second concept let the ***E***×***B*** drift sort the slow electrons stage-by-stage out from the path of the traveling electron beam. It was found through theoretical analysis and simulations, that the ***E***×***B*** concept is indeed a better choice for the implementation of an MDC for high-frequency gyrotrons. The concept has the following properties:

- ***E***×***B*** drifts sort electrons reliably. The drift can totally separate electron trajectories. In a constant electric field, the separation of the electron trajectories depends on the electric potential, rather than the magnitude of the electric field; which means that the distance between electrodes has no effect on the energy sorting. This effect provides a kind of flexibility for the MDC design.
- Energy sorting by ***E***×***B*** drifts can be very sharp, as shown in the optimized simulations. The sharp sorting makes a high collector efficiency possible. Designing an ***E***×***B*** collector with more than two stages is possible and makes sense due to (1) the sharp sorting and (2) the clear separation of trajectories.
- Secondary electrons are well handled for two reasons. First, the drift of the secondary electrons is in the same direction as the drift of the primary ones. Each generation of secondary electrons drift further away from the primary electron beam and is less possible to travel backwards in the same path of the primary electron beam. Second, the separation of trajectories makes it possible to place stages far away from each other. In the simulations, local fields and electron impinging angles can be optimized at individual stages to suppress the reflection of the secondary electrons.
- The ***E***×***B*** MDC concept could be less sensitive against the external magnetic field and manufacturing tolerances than the axisymmetric concept. The collector magnetic field up to 60 mT stabilizes and confines the electron beam.

Besides, there is already a magnetic field in the gyrotron collector. The desired ***E***×***B*** drift can make use of this existing magnetic field.

There are a lot of possibilities to create an ***E***×***B*** drift for the collection of electrons. They are systematically categorized in the chapter of ***E***×***B*** collectors (see fig. 3.1). Various design proposals from the potentially promising categories are investigated via simulations. The proposals to transform the annular beam to sheet beams (presented in sections 3.3 and 3.4) are not practical for the DEMO gyrotron, due to the large collector size and the requirement of special long coils. The proposal to split an annular beam to arc beams (presented in section 3.5) involves two baffles for the splitting. The baffles reduce the achievable efficiency and cause hotspots of power density in a high-power gyrotron. After all, the proposal to use a helical gap (section 3.6) is very promising. Its conceptual design is efficient and relatively small in size compared to other ones. The major drawback of this proposal is the back-stream current in the order of 700 mA for the 170 GHz 1 MW gyrotron. Several upgraded designs are proposed in section 3.7 for the suppression of the back-stream current, while it is possible to further reduce the collector size with the method presented in section 3.7.2. This is the best option for the realization of a proof-of-concept DEMO gyrotron MDC. Its mechanical design is proposed in section 3.9.

Outlook for the MDC Validation

This work is the initial step towards a highly efficient MDC for the DEMO gyrotron in long-pulse and CW operations. The development towards the world first MDC prototype for gyrotron is continuing. The principle of the finally proposed MDC will be experimentally validated for a coaxial gyrotron. Beside the verification of the expected efficiency, there are some effects and parameters which need careful investigations. They are beyond the capabilities of the currently available simulation tools. In particular, the following points should be checked and validated in the future simulations and experiments:

- **Influence of the back-stream current**
  Theoretically, there is still a back-stream current of 40 mA in the collector simulation. In [71], a body current of 10 mA is measured, although the simulation predicts a theoretical back-stream current of 640 mA, as the most back-streamed electrons are accelerated in the interaction and are recollected at the collector. The same physics might also occur in the

proposed MDC. However, the behavior of the back-stream electrons cannot be simulated, thus, it should be examined experimentally.

- **Power load density and beam sweeping mechanism at the last stage**
  The power load density is not well implemented in the three-dimensional simulation software used during this work. Therefore, the theoretical power load should be verified using a better implemented code. Besides, the power per area is not a very smooth function, since the (macro) electrons impact the surface discretely. The theoretical number depends on the interpolation and the fragmented area for the power counting.

  Depending on the depression voltages, the load at the first stage could be under the current constraint of 500 W/cm^2 in the proposed two-stage collector. Anyhow, at the second (last) stage of the collector, the peak local power load density will exceed the 500 W/cm^2 constraint even for a 1 MW gyrotron. For the 2 MW DEMO gyrotron, beam sweeping should be applied.

  There is a region at the end of the second stage reserved for the beam sweeping. Since the second stage of the MDC has a higher depression voltage than that of an SDC, the total power (and also the power density) at the second stage of the MDC is lower than that in an SDC with the same inner radius. The demand of the sweeping at the second MDC stage can be less critical than that of an SDC. A preliminary estimation of the coil system shows that the beam can be locally swept by a solenoid with a reasonable size and power at the second stage. However, due to the lack of the possibility for a precise evaluation of the power load density, the feasibility of a sweeping system should be further investigated.

- **Influence of secondary electrons**
  The influence of secondary electrons to the MDC should be experimentally verified, since the simulations are not able to sample the large number of secondary electrons. In case there is a large deviation between the simulation and the reality, the model and the implementation of the secondary electron emission in the simulation software should be calibrated with the experimental data.

A Theoretical Supplements

A.1 Dynamic Programming Algorithm for the Estimation of the Maximal Collector Efficiency

In this appendix, the best method for the calculation of theoretically maximal collector efficiency and optimal deceleration voltages are presented. The calculation assumes an ideal deceleration, in that a depression voltage could decelerate an electron until zero kinetic energy. The optimal deceleration voltages calculated from this process will be used for the preliminary depression voltages for the later realistic simulation.

The energy distribution (spectrum) of a gyrotron spent electron beam (see fig. 1.9) can be more peculiar than that of a TWT [108]. It is found in this work, that the finding of the theoretical optimal MDC depression voltages belongs to the Dynamic Programming (DP) problems [51].

According to the definition of η_{col} from eq. (1.7) and its denominator given by eq. (1.27), the only unknown parameter, which should be maximized is the recovered power P_{rec}.

Assuming that an MDC has N depression voltages

$$U_{\text{d}1} < U_{\text{d}2} < \cdots < U_{\text{d}N} \, ,$$

the ideally recovered power for an ideal deceleration is

$$P_{\text{rec}} = e\, I_{\text{b}} \sum_{i=1}^{N} U_{\text{d}i} \int_{U_{\text{d}i}}^{U_{\text{d}i+1}} f(e \cdot U_{\text{d}})\, \mathrm{d}U_{\text{d}} \, , \tag{A.1}$$

where $f(E_{\text{kin}})$ is the energy spectrum of the electron beam. For convenience, let

$$\begin{aligned} P(u_1, u_2) &= e\, I_{\text{b}}\, u_1 \int_{u_1}^{u_2} f(e \cdot U_{\text{d}})\, \mathrm{d}U_{\text{d}} \\ &= e\, I_{\text{b}}\, u_1 \left(F(e \cdot u_2) - F(e \cdot u_1)\right) , \end{aligned} \tag{A.2}$$

to be the power collected by the stage of depression voltage u_1, where $F(E_{\mathrm{kin}})$ is the cumulative distribution function. A table of the function $P(u_1, u_2)$ can be pre-calculated, in order to save the computation time. After the introduction of this function, the recovered power of an N-stage collector is

$$P_{\mathrm{rec}} = \sum_{i=1}^{N} P(U_{\mathrm{d}i}, U_{\mathrm{d}i+1}) . \tag{A.3}$$

A harmless "virtual" stage (the $N + 1$'th stage) can be assumed for the upper boundary of the depression voltages

$$U_{\mathrm{d}N+1} = \infty . \tag{A.4}$$

The so-called value functions $G_n(u)$ for the DP is defined as the maximum recoverable power with n stages for the interval of depression voltage $[u, \infty)$

$$G_n(u) = \begin{cases} \max\limits_{v>u} \left\{ P(u,v) + G_{n-1}(v) \right\} & n > 1 \\ P(u, \infty) & n = 1 . \end{cases} \tag{A.5}$$

This calculation subdivides an n-stage collector problem $G_n(u)$ into a series of $n - 1$ stage problems $G_{n-1}(v)$, which can be effectively solved in the same way.

Finally, the maximum achievable collector efficiency of an N-stage depressed collector is:

$$\max\left\{\eta_{\mathrm{col}}\right\} = \frac{G_N(U_{\mathrm{d}1})}{P_{\mathrm{beam}}} , \tag{A.6}$$

where $U_{\mathrm{d}1}$ is the first depression voltage, which should be less than the minimum value of the spent electron kinetic energy in eV. In this way, the theoretically maximal collector efficiency and the optimal depression voltages can be solved in $\mathcal{O}(m^2 n)$ time, in comparison to a naive enumeration, which would have $\mathcal{O}(m^n)$ time complexity, where m is the number of discretization on the E_{kin} axis of fig. 1.9 on page 21.

A.2 Evaluation of Magnetic Field and Vector Potential Induced by Electromagnetic Coils

In order to keep the magnetic system simple, no magnetic material ($\mu_r \neq 1$) is considered in the physical modeling and simulation. It was found that the magnetostatic solver currently implemented in CST has some issues, such as

- CST needs approximately 1 m distance before its open boundary condition in order to let the result converge to the expected value (might be a bug of CST), which causes the simulation volume to be huge.
- Sharing the mesh grids between different types of simulations was not completely implemented until the very recent version.

Therefore, the magnetic field for CST particle simulations is preferred to be imported from external numerical quadratures. The magnetic flux density $\boldsymbol{B}$ and the magnetic vector potential $\boldsymbol{A}$ of axisymmetric coils are calculated via integration of the coil currents. The $\boldsymbol{B}$ field is an input for CST trajectory simulations, while A_θ is mainly used in the calculation of the effective potentials (presented in section 2.4.2). The required integrals of the currents are analyzed in this appendix. The most effective methods to evaluate them are proposed.

A.2.1 Thin Current Loop

A constant current with an infinitesimal cross section forms a circular loop with a certain radius $r > 0$, see fig. A.1. There is a probe at radius (height) $h \geq 0$. The axial distance between the loop and the probe is z.

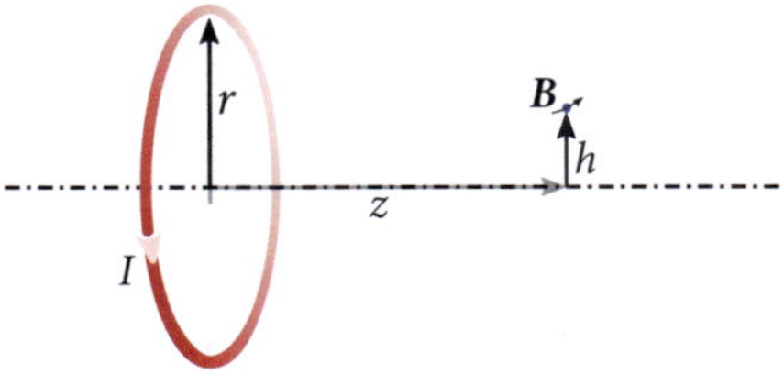

Fig. A.1: Sketch of a current loop

Magnetic Flux Density

For certain r, h and z, the magnetic flux density is calculated from the Biot-Savart integral

$$B_z = \frac{I\,\mu_0}{2\pi} \int_0^\pi \frac{r^2 - r\,h\,\cos\theta}{\sqrt{f}^3} \mathrm{d}\theta\,, \tag{A.7}$$

$$B_r = \frac{I\,\mu_0}{2\pi} \int_0^\pi \frac{-r\,z\,\cos\theta}{\sqrt{f}^3} \mathrm{d}\theta\,, \tag{A.8}$$

where

$$\begin{aligned} f &= z^2 + r^2 \sin^2\theta + (h - r\cos\theta)^2 \\ &= h^2 + r^2 + z^2 - 2hr\cos\theta \end{aligned}\,. \tag{A.9}$$

Since $|2\,h\,r\cos\theta| \le |2\,h\,r|$, the term f cannot be negative:

$$f \ge z^2 + (|h| - |r|)^2 \ge 0\,. \tag{A.10}$$

The term f is zero, only when the probe is on the path of the current loop while $\cos\theta = 1$. This position is exactly at the boundary of both integrals, thus it can be specially handled.

Magnetic Vector Potential

The magnetic vector potential $\boldsymbol{A}$ only has the azimuthal component

$$A_\theta = \frac{I\,\mu_0}{2\pi} \int_0^\pi g\,\mathrm{d}\theta\,, \tag{A.11}$$

where

$$g = \frac{r\,\cos\theta}{\sqrt{f}}\,. \tag{A.12}$$

A.2.2 Solenoid

The evaluation of the magnetic flux density and the vector potential of a solenoid involves triple integrals. Depending on the order of the integrations, one or two levels of the triple integrals have analytical solutions expressed in elementary functions. The definitions of parameters are the same as in appendix A.2.1, except that J is the current density.

Magnetic Flux Density

The original triple integrals for B_z and B_r are

$$B_z = \frac{J\,\mu_0}{2\pi} \iint \mathrm{d}z\,\mathrm{d}r \int_0^{\pi} \frac{r^2 - r\,h\,\cos\theta}{\sqrt{f}^{3}} \mathrm{d}\theta\,, \tag{A.13}$$

$$B_r = \frac{J\,\mu_0}{2\pi} \iint \mathrm{d}z\,\mathrm{d}r \int_0^{\pi} \frac{-r\,z\,\cos\theta}{\sqrt{f}^{3}} \mathrm{d}\theta\,, \tag{A.14}$$

Longitudinal Component If the most inner level of the integral is over z or r, it has an analytically solution. The indefinite integral over z is

$$B_z = \frac{J\,\mu_0}{2\pi} \iint \left[\frac{(r^2 - r\,h\,\cos\theta)\,z}{(f - z^2)\sqrt{f}} \right]_{z_1}^{z_2} \mathrm{d}\theta\,\mathrm{d}r\ . \tag{A.15}$$

However, this order is not preferred, since the denominator $(f - z^2)$ may be zero, which creates a pole singularity for the evaluation of B_z. On the other hand, if r is the most inner integration variable,

$$B_z = \frac{J\,\mu_0}{2\pi} \iint \left[\ln\left(\sqrt{f} + r - h\,\cos\theta\right) - \frac{r}{\sqrt{f}} \right]_{r_1}^{r_2} \mathrm{d}\theta\,\mathrm{d}z\ . \tag{A.16}$$

The outer integrals (for z and θ) have to be evaluated numerically.

Radial Component Two levels of the triple integration eq. (A.14) can be calculated analytically such that

$$B_r = \frac{J\mu_0}{2\pi} \int \left[\ln\left(\sqrt{f} + r - h\cos\theta\right) h\cos^2\theta + \sqrt{f}\cos\theta \right]_{z_1, r_1}^{z_2, r_2} \mathrm{d}\theta\,, \tag{A.17}$$

The integration over θ has to be calculated numerically.

Handling Singularity When $h > r$, it may happen that

$$\sqrt{f} + r - h\cos\theta < 0\,. \tag{A.18}$$

In this case, both eqs. (A.16) and (A.17) are not well-defined. If this occurs, negating h and r altogether such that $h' = -h$, $r' = -r$ will solve this issue.

A probe inside the coil can be singular, when $\sqrt{f} + r - h\cos\theta = 0$. Splitting the integral at this position, the singular point becomes the common end of the resulted sub-intervals. A few numerical quadrature methods (e.g. Gaussian quadrature) may handle such infinity at the end of an interval very well, otherwise integral transformations like the IMT method [109] can be applied, as shown in the following description.

In order to evaluate

$$Y = \int_a^b u(x)\,\mathrm{d}x\,, \tag{A.19}$$

where $u(x)$ is infinite at $x = a$ or $x = b$; let

$$\psi = \exp\left(\frac{-c}{1-t^2}\right)\,, \tag{A.20}$$

$$\phi = a + \frac{b-a}{\gamma}\int_{-1}^{t} \psi(u)\,\mathrm{d}u \tag{A.21}$$

where

$$\gamma = \int_{-1}^{1} \psi(u)\,\mathrm{d}u\,.$$

Then

$$\phi'(t) = \frac{b-a}{\gamma}\psi(t)\,. \tag{A.22}$$

Finally,

$$Y = \int_{-1}^{1} u\big(\phi(t)\big)\, \phi'(t)\, \mathrm{d}t \tag{A.23}$$

is equivalent to the original integral in (A.19), whereas the integrand and its derivatives vanish at both ends of the interval.

Magnetic Vector Potential

The A_θ of a solenoid is

$$A_\theta = \frac{J\,\mu_0}{2\pi} \iint \mathrm{d}z\,\mathrm{d}r \int_0^\pi g\ \mathrm{d}\theta\,, \tag{A.24}$$

i.e.

$$A_\theta = \frac{J\,\mu_0}{2\pi} \iint \mathrm{d}z\,\mathrm{d}r \int_0^\pi \frac{r\ \cos\theta}{\sqrt{z^2 + r^2\sin^2 + \theta(h - r\cos)\theta^2}}\ \mathrm{d}\theta\,. \tag{A.25}$$

Analytically Solvable Part of the Triple Integral Similar to the triple integral for B_z, there is an analytical solution for either r or z.

The indefinite integral over r yields

$$\int g\,\mathrm{d}r = \cos\theta \left[h\ \cos\theta\ \ln(r - h\cos\theta + \sqrt{f}) + \sqrt{f}\right]. \tag{A.26}$$

It is more complicate than the integration over z. Moreover, there is a singularity when $h = r\ \cos\theta$. Therefore, this approach is not very appropriate.

If the most inner level of the triple integral is the $\int \mathrm{d}z$, then

$$\int g\,\mathrm{d}z = r\ \cos\theta\ \ln(z + \sqrt{f})\,. \tag{A.27}$$

This is a better approach. The rest two levels of the integration should be performed numerically.

Singularities There can be two kinds of singularities appearing during the numerical quadratures.

The first kind is at some positions outside of coils when the integration $\int_{z_1}^{z_2}$ have negative boundaries, i.e. z_1 and z_2 are both negative. It causes $z + f^{1/2} = 0$, which yields a singularity of the type $\ln(0/0)$. This can be solved by mirroring the integral interval: instead of $\int_{z_1}^{z_2}$, the integration over $\int_{-z_2}^{-z_1}$ is to be calculated based on the symmetry.

The second kind is a probe within the coil. $\ln(z + \sqrt{f}) = \ln(0)$ is caused by $r = h$ at the point where $\cos\theta = 1$. When this happens, only $1/|z|$ remains in the integrand of eq. (A.27) (the nominator $r \cos\theta$ is a constant to this integral). Equation (A.27) becomes

$$\int |z|^{-1}\,\mathrm{d}z = \operatorname{sgn}(z)\ \ln(|z|) \tag{A.28}$$

Final Expression Finally, the expression for A_θ is

$$A_\theta = \frac{J\,\mu_0}{2\pi} \int_0^{\pi} \int_{r_1}^{r_2} \Big[r\ \cos\theta\ \ln(z + \sqrt{f}) \Big]_{z_1}^{z_2} \mathrm{d}r\,\mathrm{d}\theta \tag{A.29}$$

where $r_{1,2}$ are the inner and outer radius of the coil, $z_{1,2}$ are the axial distances from the probe to both ends of the solenoid. Integrals over r and θ should be evaluated numerically. There is a singularity at $z = 0$, $h = r$ when $\cos\theta = 1$, the total interval can be split at this point or transformed via IMT, as in the evaluation of $\boldsymbol{B}$.

A.2.3 Current Segment

Composing the magnetic field in fig. 3.6 requires the magnetic field from a straight current segment. For a current segment parallel to z (visualized in fig. A.2), the magnitude of the magnetic flux density at (x_0, y_0, z_0) is

$$
\begin{aligned}
B &= \frac{\mu_0}{4\pi} I \int_{z_1}^{z_2} \frac{r}{\sqrt{r^2 + (z_0 - z)^2}^3} \, \mathrm{d}z \\
&= \frac{\mu_0}{4\pi} I \left[\frac{z - z_0}{r\sqrt{r^2 + (z - z_0)^2}} \right]_{z=z_1}^{z=z_2} ,
\end{aligned}
\tag{A.30}
$$

where

$$
r = \sqrt{(x - x_0)^2 + (y - y_0)^2} \, . \tag{A.31}
$$

The normalized vector of the field is

$$
\hat{\boldsymbol{B}} = \frac{1}{r}(y - y_0, x_0 - x, 0)^{\mathrm{T}} \, . \tag{A.32}
$$

Finally, the flux density is

$$
\begin{aligned}
\boldsymbol{B} = B\hat{\boldsymbol{B}} = &\begin{pmatrix} y - y_0 \\ x_0 - x \\ 0 \end{pmatrix} \frac{\mu_0}{4\pi} I \frac{1}{(x - x_0)^2 + (y - y_0)^2} \\
&\left[\frac{z - z_0}{\sqrt{(x - x_0)^2 + (y - y_0)^2 + (z - z_0)^2}} \right]_{z=z_1}^{z=z_2} .
\end{aligned}
\tag{A.33}
$$

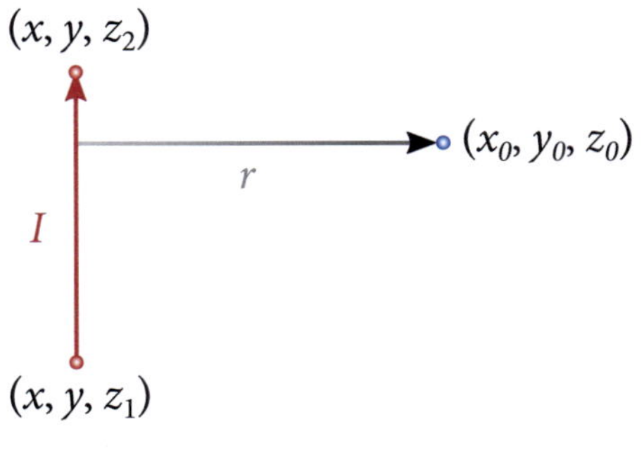

Fig. A.2: Current segment I induces a field at (x_0, y_0, z_0)

A.3 An Ideal Smooth Transition of Magnetic Field

To create an example of a non-adiabatic transition, a smooth magnetic field is analytically defined in this appendix. The non-adiabatic transition takes place in the interval $z \in [-z_0, z_0]$, while the regions before and after the transition have two (different) homogeneous magnetic fields.

To ensure the smoothness of the transition, a space $\mathcal{K}_\iota^n(\mathbb{R})$ with $\iota > 0$ is defined. It is a subspace of $\mathcal{C}^n(\mathbb{R})$ with an additional limitation that $\forall f_\iota \in \mathcal{K}_\iota^n$

$$f_\iota(x) = 0 \text{ for } |x| > \iota\,. \tag{A.34}$$

For instances

$$f_\iota(x) = \begin{cases} 1 - \cos\left(\left[\frac{x}{\iota} - 1\right]\pi\right) & |x| < \iota \\ 0 & \text{otherwise} \end{cases} \in \mathcal{K}_\iota^1\,, \tag{A.35a}$$

$$\text{for } n \geq 0 \quad f_\iota(x) = \begin{cases} \left(1 - \frac{1}{\iota^2}x^2\right)^{n+1} & |x| < \iota \\ 0 & \text{otherwise} \end{cases} \in \mathcal{K}_\iota^n\,, \tag{A.35b}$$

$$f_\iota(x) = \begin{cases} \exp\left(\frac{x^2}{x^2 - \iota^2}\right) & |x| < \iota \\ 0 & \text{otherwise} \end{cases} \in \mathcal{K}_\iota^\infty\,. \tag{A.35c}$$

The model of the magnetic field has to meet the following constraints:

- Gauß's law of an axisymmetric magnetic field

$$B_\theta = 0 \text{ and } \boldsymbol{\nabla} \cdot \boldsymbol{B} = 0 \quad \Leftrightarrow \quad \frac{1}{r}\frac{\partial r B_r}{\partial r} + \frac{\partial B_z}{\partial z} = 0\,. \tag{A.36a}$$

- The homogeneity before and after the transition

$$\boldsymbol{\nabla} B_z = 0 \quad \text{for} \quad |z| > z_0\,. \tag{A.36b}$$

- Outside of the transition region, the radial component of the magnetic field vanishes

$$B_r(z,r) = 0 \quad \text{for} \quad |z| > z_0 \,. \tag{A.36c}$$

- In addition,

$$B_r(z, r = 0) = 0 \,. \tag{A.36d}$$

To synthesize a smooth magnetic field, two functions $\varphi \in \mathcal{K}^2_{z_0}$ and $\psi \in \mathcal{K}^1_{z_0}$ will be introduced as follows. The non-unique definition of the field is based on the variable separation

$$B_z(z,r) = \rho(r)\frac{\partial\varphi(z)}{\partial z} + \zeta(z) \tag{A.37a}$$

with

$$\zeta(z) := \zeta_1 \int_{-\infty}^{z} \psi(z)\,\mathrm{d}z \; + \; \zeta_0 \tag{A.37b}$$

where $\rho(r)$ only depends on r, $\varphi(z)$ and $\psi(z)$ only on z. This magnetic field fulfills eq. (A.36b) since

$$\frac{\partial B_z}{\partial z} = \rho(r)\frac{\partial\varphi(z)}{\partial z} + \zeta_1\psi(z) \tag{A.38a}$$

and

$$\frac{\partial B_z}{\partial r} = \varphi(z)\frac{\partial\rho(r)}{\partial r} \tag{A.38b}$$

vanish outside of the transition region.

Substituting eq. (A.38a) into eq. (A.36a) yields

$$rB_r = -\left[\frac{\partial\varphi(z)}{\partial z}\int_0^r r\rho(r)\,\mathrm{d}r \; + \; \frac{1}{2}r^2\zeta_1\psi(z)\right] \,. \tag{A.39}$$

All constraints are so far automatically fulfilled for a proper $\rho(r)$.

The simplest choice of $\rho(r)$ is zero. In that case,

$$B_z(z) = \zeta_1 \int_{-\infty}^{z} \psi(z)\,\mathrm{d}z \; + \; \zeta_0 \,, \tag{A.40a}$$

$$B_r(z,r) = -\frac{1}{2} r \zeta_1 \psi(z) \,. \tag{A.40b}$$

For a polynomial $\psi(z) \in \mathcal{K}^1_{z_0}$ that

$$\psi(z) = \begin{cases} \left[1 - \left(\frac{z}{z_0}\right)^2\right]^2 & |z| < z_0 \\ 0 & \text{otherwise} \end{cases} \,, \tag{A.41}$$

eq. (A.40) becomes eq. (2.1).

A.4 Derivation of the Drift Distance for Proposal IV

The goal of this analysis is to find out the required minimal drift distance in order to place the conic wall with a reasonable distance to the electron beam. The unknown parameters in eq. (3.33) are the angle ϑ and the total local magnetic field B. Assuming that the electric field and the magnetic field near the helix are not very different from that at the location of electron beam, the values of $\cos\vartheta$ and B along the helix are calculated as follows.

The function of a helical curve $\boldsymbol{h}$ from the axial position $z = z_0$ to $z = z_0 + L$ has the form

$$\boldsymbol{h}(z) = (h_x, h_y, z) \,. \tag{A.42}$$

For convenience, the relative axial position l is employed instead of the absolute coordinate z:

$$l := z - z_0 \,, \tag{A.43}$$

The definition of the curve $\boldsymbol{h}(l)$ is not unique. The simplest expression requires two auxiliary variables, which vary linearly with l:

$$r = r_0 + \rho\, l \,, \tag{A.44a}$$

$$\theta = 2\pi \frac{l}{L} , \tag{A.44b}$$

where L is the total axial height of the helix, r_0 is the initial radius and $\rho \geq 0$ as marked in fig. 3.24. The x and y components of $\boldsymbol{h}$ are

$$h_x = r \cos\theta = (r_0 + \rho\, l) \cos 2\pi \frac{l}{L} , \tag{A.45a}$$

$$h_y = r \sin\theta = (r_0 + \rho\, l) \sin 2\pi \frac{l}{L} . \tag{A.45b}$$

Accordingly, the growing direction of the helix is

$$\boldsymbol{h}' = \frac{\mathrm{d}\boldsymbol{h}}{\mathrm{d}l} = \left(\frac{\mathrm{d}h_x}{\mathrm{d}l}, \frac{\mathrm{d}h_y}{\mathrm{d}l}, 1 \right) . \tag{A.46}$$

Another auxiliary vector is the normal vector of the conical surface, where the helix is located. It has the direction

$$\boldsymbol{\nu} = (\cos\theta, \sin\theta, -\rho) , \tag{A.47}$$

which is marked in fig. 3.24.

Since the local magnetic field is approximately parallel to the conical surface and free of azimuthal component, it can be decomposed as

$$B_z = |B| \frac{1}{\sqrt{1 + \rho^2}} , \tag{A.48a}$$

$$B_r = |B| \frac{\rho}{\sqrt{1 + \rho^2}} = \rho\, B_z , \tag{A.48b}$$

where ρ is the steepness of the radius along z, see eq. (A.44a). For the calculation of $|B|$ it is reasonable to assume that B_z is nearly constant over r ($\pm 2\,\%$ error in the field involved in section 3.6). According to the flux conservation,

$$B_z = \frac{\psi}{\pi\, r^2} , \tag{A.49}$$

where ψ is the magnetic flux enclosed within a circle of radius r. The flux density at any point of the helix is

$$\boldsymbol{B} = \frac{\psi}{\pi\, r^2} (\rho \cos\theta, \rho \sin\theta, 1)^{\mathrm{T}} . \tag{A.50}$$

In order to calculate the angle ϑ in eq. (3.33), the direction of the local electric field is required. A vector $\mathscr{E}$ has the same direction as the electric field $\boldsymbol{E}$ at the helical gap, i.e.,

$$\mathscr{E} = K\boldsymbol{E}\,, \tag{A.51}$$

where $K > 0$. It fulfills the following three conditions:

1. $\mathscr{E}$ is tangential to the surface of the cone, where the helix is located

$$\boldsymbol{\nu} \cdot \mathscr{E} = 0\,. \tag{A.52a}$$

2. $\mathscr{E}$ is perpendicular to the helical electrode surface of the first stage, where the slow electrons are supposed to be collected, which requires $\mathscr{E}$ to be perpendicular to the growing direction of the helix

$$\boldsymbol{h}' \cdot \mathscr{E} = 0\,. \tag{A.52b}$$

3. $\mathscr{E}$ points towards the second stage

$$\mathscr{E}_z > 0\,. \tag{A.52c}$$

One solution which can fit these three criteria is

$$\mathscr{E} = \begin{pmatrix} 2\pi r\rho\cos\theta + L(1+\rho^2)\sin\theta \\ 2\pi r\rho\sin\theta - L(1+\rho^2)\cos\theta \\ 2\pi r \end{pmatrix}. \tag{A.53}$$

Finally, the scalar product of eqs. (A.50) and (A.53) gives the angle between the electric field and the magnetic field

$$\cos\vartheta = \frac{\mathscr{E}\cdot\boldsymbol{B}}{|\mathscr{E}||B|}\,. \tag{A.54}$$

The drift distance D is calculated by substituting eqs. (A.50) and (A.54) into eq. (3.33). This will result in a very lengthy equation. The elegance of this model is that this lengthy expression has an equivalent simple form:

$$D = \frac{m_e v_0 L}{q\psi} r = \frac{m_e v_0 L\rho}{q\psi} l + \frac{m_e v_0 L r_0}{q\psi}\,. \tag{A.55}$$

B Additional Collector Design Proposals

B.1 An Existing *E*×*B*-Type of Design Proposal Using Azimuthal Magnetic Field

E*×*B MDCs using an azimuthal magnetic field component B_θ were proposed at meantime by Peter the Great St. Petersburg Polytechnic University [89–92].

The advantage of such a proposal is that the shapes of the outer electrodes can be more regular than the case with E_θ, such that the technologies for conventional MDCs (not only for gyrotrons) may be adapted. However, the major challenge is that there should be a current with sufficient intensity enclosed within the electron beam according to Ampère's law. How to bury this current is the first issue to target. Figure B.1 shows a collector coil system which creates the required components of magnetic field. In addition to the solenoids (red), which tune the longitudinal component of the magnetic field, there are numerous longitudinal coils (gold) creating the B_θ component. The longitudinal coils break the annular shape of electron beam at two azimuthal angles.

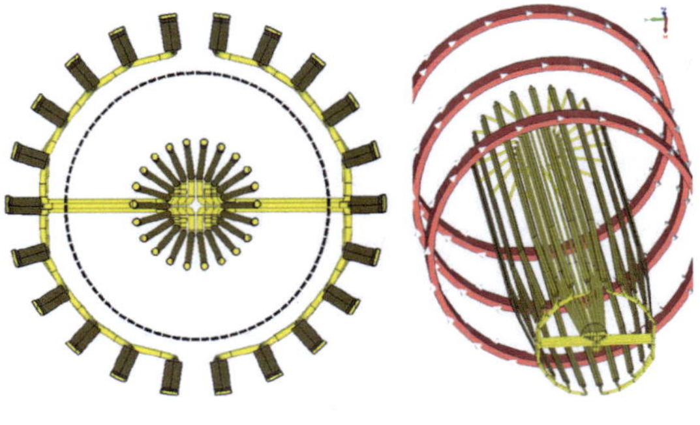

Fig. B.1: Proposal of a coil scheme for an ***E*×*B*** MDC using B_θ [106]

Altogether, more coils than the sheet-beam approach (section 3.3) are required. The massive coils burden the total weight of an MDC. Besides, according to fig. 3.5, the transverse magnetic field should match the axial component in order to effectively create the drift. The investigations on this approach are excluded from the presented work, because it seems to be more challenging than the proposals presented in chapter 3.

B.2 A Different Kind of Axisymmetric Design Proposal Using Electric Field to Sort Electrons

It has been observed in experiments [72], that if the electron beam is close (approximately one Larmor radius) to the wall, the trapped (back-and-forth reflected) electrons can be collected at that place due to diffusion. It has been reported in that experiment, that 10 % of the beam current can be collected in this way, whereas the output power is not affected. The single-stage depressed collector becomes a two-stage collector, with one of the stages depressed.

There could be issues for the design of such an MDC. First, the high voltage power supply for the acceleration voltage are usually not designed to endure the high current. Second, the magnetic field has to be very accurate. Anyhow, the diffusion process cannot be simulated with the currently available simulation programs yet.

Inspired by this experiment, another type of axisymmetric MDC is conceptually investigated. Distinct from the MDC type in section 2.4, where there was a transition of magnetic field; this type of energy sorting uses electric field. Figure B.2 shows the conceptual simulation of such a two-stage collector for a coaxial gyrotron. The profile of the first stage has a wedge form, whose inner side is parallel to the magnetic flux surface. The sorting can happen in a relatively high magnetic field. For instance, the field in region $z \in [1.6, 1.7]$ m is higher than 40 mT, which is four times of the field in section 2.4. Therefore, no additional coil is needed for the demagnetization at all.

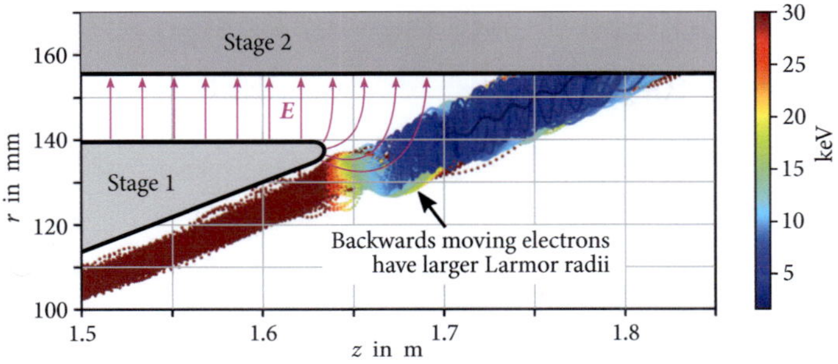

Fig. B.2: Axisymmetric MDC based on the sudden change of electric field

The forward moving electron beam has Larmor radius much less than 2 mm and the electron beam is at least 5 mm away from the wedge. As the beam leaves the region beneath the wedge, electrons are decelerated. During the deceleration, the electric field affects the electron trajectories inside one cyclotron period, so that the electrons gain transverse speed. The slow electrons cannot reach the second stage. They will turn back (reflected). Most of the reflected electrons have large Larmor radii (5 to 10 mm). Such that these electrons are finally collected at the wedge. Some reflected electrons with insufficient Larmor radius are trapped by the compressing magnetic field and then they enter the deceleration region between the two stages again. This time their Larmor radii probably increase, such that they can also be collected at the first stage. 200 mA of the injected 70 A beam current are back-streamed to the cavity. These back-stream electrons may participate the interaction and re-enter the collector to be collected like the other electrons. Diffusion is not considered in the simulation and is nevertheless not possible to take into account. The diffusion may improve the collector performance as in the experiment.

The major advantage of this design is its simplicity. Such a two-stage depressed collector may be directly prototyped within an SDC and does not need any additional coil in a short-pulsed experiment. A longitudinal sweeping is applicable at the second stage. However, the sorting mechanism does not ensure a clear energy sorting, and the diffusion, which according to the experiment is important, cannot be considered in the simulation. Therefore, this design may not be the best candidate for the first gyrotron MDC prototype due to the potential risks.

B.3 An Additional *E×B*-Type of Design Proposal Using Inverted Helical Electrodes

Another conceptual design of the ***E***×***B*** MDC using helical electrodes is presented in this appendix. There are two highlights in this design

1. It has one layer less than the design presented in section 3.9.
2. Instead of cooling helical surfaces like fig. 3.40, mainly a disk is needed to be cooled at the first stage.

Figure B.3 shows the geometry of the ***E***×***B*** MDC design, which is cut in the middle. Figures B.4a and B.4b are its perspective top views. The outer vacuum envelope is electrically connected to the mirror box. They have the potential of the second stage. In the mirror box, acceleration structures [49] have to be applied in order to ensure that all electrons can enter the collector. The upper helical structures at the second stage are fixed at the end of vacuum envelope, while the lower helical structures are hung from top, where there is the insulation between the two stages. Both of them are thin plates and no electron should theoretically impact their surfaces. The helices curl in the opposite direction as in section 3.6, therefore, electrons drift inwards and penetrate the helical gaps.

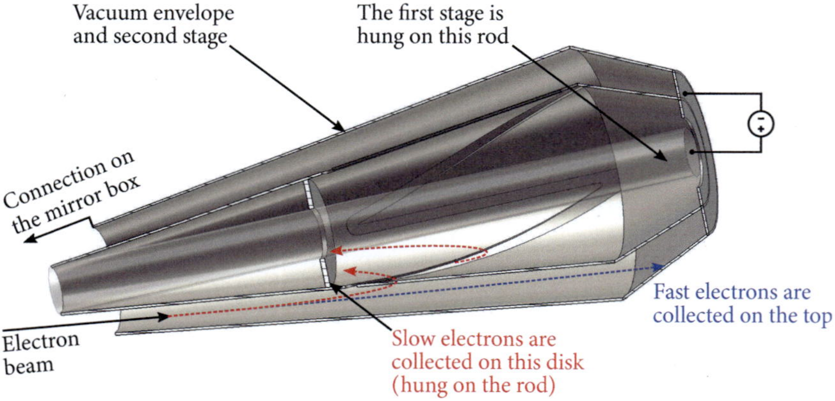

Fig. B.3: Geometry of the MDC with inward drift

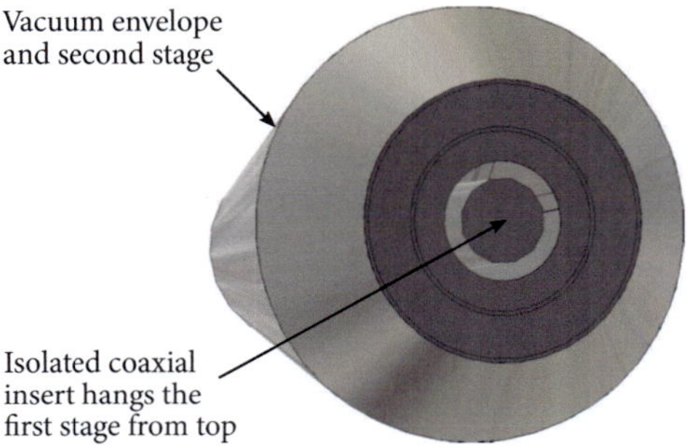

(a) Geometry

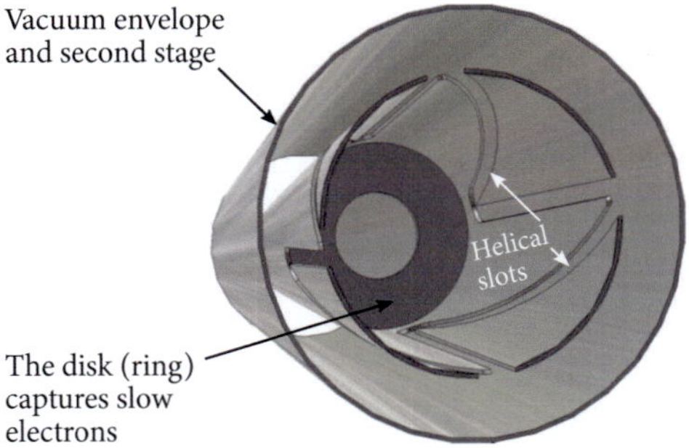

(b) Geometry with opened top

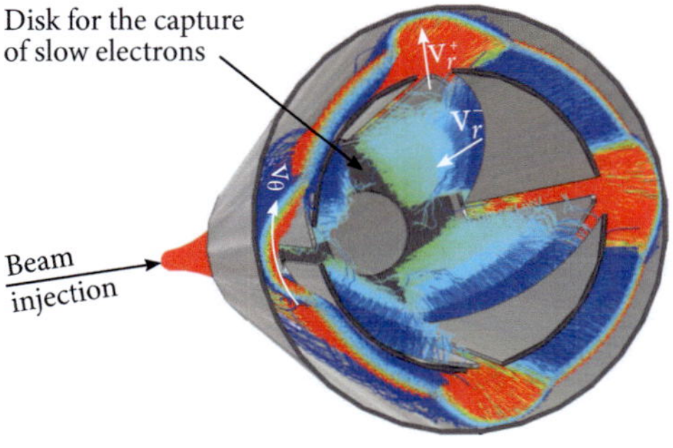

(c) Trajectories

Fig. B.4: Top view of the MDC with inward drift

Figure B.4c shows the electron trajectories in the stationary state. There is an inwards drift velocity v_r^- pushing slow electrons through the helical gaps. Due to the longitudinal electric field, electrons inside the inner conic layer are accelerated backwards to the disk.

There are four helices distributed at four azimuthal angles. This also means that at four azimuthal positions, there are undesirable outwards drifts (see the v_r^+ in fig. B.4c). The outer and inner conic layers create a radial electric field, hence, there is an additional azimuthal drift preventing a certain number of electrons at those azimuthal angles from being reflected, as explained in section 3.7.1.

Simulations show a similar collector efficiency as the designs with ordinary helical electrodes in section 3.6. There is a back-stream current of 100 mA from a 70 A spent electron beam, while considering secondary electrons, the back-stream current increases to 300 mA. However, this design is not as perfect (considering the electron back-streaming) and flexible (in the choice of depression voltages) as the design presented section 3.9. Besides, the functionality of the acceleration [49] structure is still under verification. This proposal has to be further optimized and investigated.

Bibliography

[1] V. Erckmann and U. Gasparino. “Electron cyclotron resonance heating and current drive in toroidal fusion plasmas”. In: *Plasma Physics and Controlled Fusion* 36.12 (1994), pp. 1869–1962. ISSN: 0741-3335. DOI: 10/fdn6kb (cit. on p. 1).

[2] S. Marsen, Y. Corre, H. P. Laqua, V. Moncada, D. Moseev, et al. “First results from protective ECRH diagnostics for Wendelstein 7-X”. In: *Nuclear Fusion* 57.8 (June 2017), p. 086014. ISSN: 0029-5515. DOI: 10/gdvmst (cit. on p. 1).

[3] R. Aymar, V. A. Chuyanov, M. Huguet, and Y. Shimomura. “Overview of ITER-FEAT - The future international burning plasma experiment”. In: *Nuclear Fusion* 41.10 (2001), pp. 1301–1310. ISSN: 0029-5515. DOI: 10/djkk6z (cit. on p. 1).

[4] J. P. Freidberg. *Plasma Physics and Fusion Energy*. Cambridge University Press, 2008. ISBN: 978-1-139-46215-0 (cit. on p. 1).

[5] K. Sakamoto, A. Kasugai, K. Takahashi, R. Minami, N. Kobayashi, et al. “Achievement of robust high-efficiency 1-MW oscillation in the hard-self-excitation region by a 170-GHz continuous-wave gyrotron”. In: *Nature Physics* 3.6 (Apr. 2007), pp. 411–414. DOI: 10/dbv36x (cit. on pp. 1, 9–10).

[6] M. E. Read, W. G. Lawson, A. J. Dudas, and A. Singh. “Depressed collectors for high-power gyrotrons”. In: *IEEE Trans. Electron Devices* 37.6 (June 1990), pp. 1579–1589. ISSN: 0018-9383. DOI: 10/dnr2gq (cit. on pp. 2, 35, 42, 44).

[7] A. Singh, V. L. Granatstein, G. Hazel, G. Saraph, J. M. Cooperstein, et al. “A depressed collector system for a quasi-optical gyrotron with precisely controlled magnetic flux lines”. In: *International Journal of Infrared and

Millimeter Waves 12 (Apr. 1991), pp. 323–334. DOI: 10/b2744p (cit. on pp. 2, 35, 66, 77).

[8] A. Singh, G. Hazel, V. L. Granatstein, and G. Saraph. "Magnetic field profiles in depressed collector region for small-orbit gyrotrons with axial or radially extracted spent beam". In: *International Journal of Electronics* 72.5-6 (May 1992), pp. 1153–1163. ISSN: 0020-7217. DOI: 10/dfwkw4 (cit. on pp. 2, 35, 42, 66, 77).

[9] A. Singh, D. S. Weile, S. Rajapatirana, and V. L. Granatstein. "Integrated design of depressed collectors for gyrotrons". In: *IEEE Trans. Plasma Sci.* 25.3 (June 1997), pp. 480–491. ISSN: 0093-3813. DOI: 10/d78gk3 (cit. on pp. 2, 35, 37, 65–66, 77).

[10] A. Singh, S. Rajapatirana, Y. Men, V. L. Granatstein, R. L. Ives, et al. "Design of a multistage depressed collector system for 1-MW CW gyrotrons. I. Trajectory control of primary and secondary electrons in a two-stage depressed collector". In: *IEEE Trans. Plasma Sci.* 27.2 (Apr. 1999), pp. 490–502. ISSN: 0093-3813. DOI: 10/cmjv6c (cit. on pp. 2, 19, 35–36, 44, 49, 63, 66, 69, 73, 77, 152).

[11] G. Ling, B. Piosczyk, and M. K. Thumm. "A new approach for a multi-stage depressed collector for gyrotrons". In: *IEEE Trans. Plasma Sci.* 28.3 (June 2000), pp. 606–613. ISSN: 0093-3813. DOI: 10/c26j4b (cit. on pp. 2, 35, 42, 44, 63, 66, 69, 77).

[12] M. Y. Glyavin, M. V. Morozkin, and M. I. Petelin. "Separation of energy fractions of an electron beam by a localized nonuniformity of magnetic field in the collector region of gyrodevices". In: *Radiophysics and quantum electronics* 49.10 (2006), pp. 811–815. ISSN: 0033-8443. DOI: 10/c3v7pk (cit. on pp. 2, 36–37, 44, 66).

[13] C. Wu, I. G. Pagonakis, S. Illy, G. Gantenbein, M. Thumm, et al. "Novel multistage depressed collector for high power fusion gyrotrons based on an E×B drift concept". In: *Proc. Int. Vacuum Electronics Conf. (IVEC)*. Apr. 2017, pp. 1–2. DOI: 10/gdvmr5 (cit. on pp. 2, 80).

[14] C. Wu, I. G. Pagonakis, K. A. Avramidis, G. Gantenbein, S. Illy, et al. "Gyrotron multistage depressed collector based on E×B drift concept using azimuthal electric field. I. Basic design". In: *Physics of Plasmas* 25.3 (2018), p. 033108. DOI: 10/gdvmsd (cit. on pp. 2, 19, 80, 123–125, 128–129).

[15] C. Wu, I. G. Pagonakis, D. Albert, K. A. Avramidis, G. Gantenbein, et al. "Gyrotron multistage depressed collector based on E×B drift concept using azimuthal electric field. II: Upgraded designs". In: *Physics of Plasmas* 26.1 (Jan. 2019), p. 013108. DOI: 10/czw6 (cit. on pp. 2, 80, 136, 139–141, 143).

[16] W. M. Stacey. *Fusion Plasma Physics.* Wiley VCH Verlag GmbH, Aug. 22, 2012. ISBN: 3-527-41134-8 (cit. on p. 3).

[17] J. Wesson. *Tokamaks.* OUP Oxford, Oct. 13, 2011. 832 pp. ISBN: 0-19-959223-3 (cit. on p. 3).

[18] J. Jacquinot. "Progress and prospect of true steady state operation with RF". In: *European Physical Journal Web of Conferences.* Vol. 157. European Physical Journal Web of Conferences. Oct. 2017, p. 01002. DOI: 10/gdvmsr (cit. on p. 3).

[19] G. S. Jeffrey Forshaw. *Dynamics and Relativity.* John Wiley & Sons, 2014. 344 pp. (cit. on p. 4).

[20] G. Granucci, G. Aiello, S. Alberti, K. A. Avramidis, F. Braunmüller, et al. "Conceptual design of the EU DEMO EC-system: main developments and R&D achievements". In: *Nuclear Fusion* 57.11 (Aug. 2017), p. 116009. ISSN: 0029-5515. DOI: 10/gdvmss (cit. on p. 4).

[21] F. Leuterer, M. Beckmann, H. Brinkschulte, F. Monaco, M. Münich, et al. "Experience with the ECRH system of ASDEX-Upgrade". In: *Fusion Engineering and Design* 53.1-4 (Jan. 2001), pp. 485–489. DOI: 10/fqfrht (cit. on p. 4).

[22] F. Gandini, T. S. Bigelow, B. Becket, J. B. Caughman, D. Cox, et al. "The EC H&CD transmission line for ITER". In: *Fusion Science and Technology* 59.4 (2011), pp. 709–717. DOI: 10/gdvmsp (cit. on p. 5).

[23] R. L. Ives, A. Singh, M. Mizuhara, R. Schumacher, J. Neilson, et al. "Design of a multistage depressed collector system for 1-MW CW gyrotrons. II. System consideration". In: *IEEE Trans. Plasma Sci.* 27 (Apr. 1999), pp. 503–511. DOI: 10/brpbdx (cit. on pp. 4, 44, 63, 66, 77, 133).

[24] T. Rzesnicki, F. Albajar, S. Alberti, K. A. Avramidis, W. Bin, et al. "Experimental verification of the European 1-MW, 170-GHz industrial CW prototype gyrotron for ITER". In: *Fusion Engineering and Design* 123 (2017), pp. 490–494. ISSN: 0920-3796. DOI: 10/gct9g4 (cit. on pp. 6, 46, 108).

[25] B. Piosczyk, G. Dammertz, O. Dumbrajs, S. Illy, J. Jin, et al. "A 2-MW, 170-GHz coaxial cavity gyrotron — experimental verification of the design of main components". In: *Journal of Physics: Conference Series* 25 (Jan. 2005), pp. 24–32. DOI: 10/fwk4st (cit. on p. 6).

[26] Z. C. Ioannidis, T. Rzesnicki, F. Albajar, S. Alberti, K. A. Avramidis, et al. "CW experiments with the EU 1-MW, 170-GHz industrial prototype gyrotron for ITER at KIT". In: *IEEE Trans. Electron Devices* 64.9 (Sept. 2017), pp. 3885–3892. ISSN: 0018-9383. DOI: 10/gdvmrx (cit. on pp. 8, 46, 108).

[27] M. Thumm. *State-of-the-art of high power gyro-devices and free electron masers.* Update 2014. KIT-SR 7693. KIT Scientific Publishing, 2015 (cit. on p. 9).

[28] I. G. Pagonakis, F. Albajar, S. Alberti, K. Avramidis, T. Bonicelli, et al. "Status of the development of the EU 170-GHz/1-MW/CW gyrotron". In: *Fusion Engineering and Design* 96-97 (Oct. 2015), pp. 149–154. DOI: 10/gfrh78 (cit. on p. 9).

[29] W. K. H. Panofsky. *Classical Electricity and Magnetism.* Dover Publications Inc., Jan. 26, 2005. 512 pp. ISBN: 0-486-43924-0 (cit. on p. 12).

[30] F. F. Chen. "Single-Particle Motions". In: *Introduction to Plasma Physics and Controlled Fusion*. Springer, Jan. 2016. ISBN: 978-3-319-22308-7 (cit. on p. 14).

[31] S. E. Tsimring. *Electron beams and microwave vacuum electronics*. Vol. 191. John Wiley & Sons, 2006 (cit. on p. 15).

[32] H. Goldstein, C. P. Poole, and J. L. S. Sr. *Klassische Mechanik (German Edition)*. 2012. ISBN: 3-527-66207-3 (cit. on p. 15).

[33] H. Busch. "Berechnung der Bahn von Kathodenstrahlen im axialsymmetrischen elektromagnetischen Felde". In: *Annalen der Physik* 386.25 (1926), pp. 974–993. ISSN: 0003-3804. DOI: 10/dchtj7 (cit. on p. 15).

[34] C. R. Donaldson, W. He, A. W. Cross, F. Li, A. D. R. Phelps, et al. "A cusp electron gun for millimeter wave gyrodevices". In: *Applied Physics Letters* 96.14 (Apr. 2010), p. 141501. DOI: 10/dtj3kg (cit. on p. 16).

[35] H. Bruining. *Die Sekundär-Elektronen-Emission fester Körper*. Springer Berlin Heidelberg, 1942. 140 pp. ISBN: 978-3-642-88894-6 (cit. on pp. 16, 20).

[36] M. A. Furman and M. T. Pivi. "Probabilistic model for the simulation of secondary electron emission". In: *Physical Review Special Topics Accelerators and Beams* 5.12, 124404 (Dec. 2002), p. 124404. DOI: 10/b6stj8 (cit. on pp. 17, 20, 102, 117).

[37] M. A. Furman and M. T. F. Pivi. "Erratum: Probabilistic model for the simulation of secondary electron emission". In: *Physical Review Special Topics Accelerators and Beams* 16.6, 069901 (June 2013), p. 069901. ISSN: 1098-4402. DOI: 10/gdvmr2 (cit. on pp. 17, 20, 117).

[38] S. Illy, J. Zhang, and J. Jelonnek. "Gyrotron electron gun and collector simulation with the ESRAY beam optics code". In: *International Vacuum Electronics Conference (IVEC)*. IEEE, Apr. 2015, pp. 1–2. DOI: 10/gdvqds (cit. on pp. 17, 19, 44, 75).

[39] CST Computer Simulation Technology AG. *CST Studio Suite® Manual*. 2016 (cit. on pp. 17, 19).

[40] F. Hamme, U. Becker, and P. Hammes. "Simulation of secondary electron emission with CST PARTICLE STUDIO". In: *Proceedings of ICAP*. 2006, pp. 160–163 (cit. on p. 17).

[41] H. Bruining and J. H. De Boer. "Secondary electron emission. Part I. Secondary electron emission of metals". In: *Physica* 5 (Jan. 1938), pp. 17–30. DOI: 10/bf9rpp (cit. on p. 18).

[42] R. Kollath. "A new method for the measurement of the energy distribution of secondary electrons". In: *Z. techn. Phys.* 21 (1940), p. 328 (cit. on p. 18).

[43] J. A. Dayton. "A review of the suppression of secondary electron emission from the electrodes of multistage collectors". In: *Proc. ISDEIV. 18th Int. Symp. Discharges and Electrical Insulation in Vacuum (Cat. No.98CH36073)*. Vol. 1. Aug. 1998, pp. 9–15. DOI: 10 / fv7kj6 (cit. on p. 20).

[44] V. Baglin, J. Bojko, O. Gröbner, B. Henrist, N. Hilleret, et al. "The secondary electron yield of technical materials and its variation with surface treatments". In: *7th European Conference on Particle Accelerator (EPAC)*. Vol. 1–3. Vienna, Austria, June 2000, pp. 217–221 (cit. on p. 20).

[45] G. Stupakov and M. Pivi. *Suppression of the Effective Secondary Emission Yield for a Grooved Metal Surface*. Tech. rep. CERN, 2004. DOI: 10/fhbpc4 (cit. on pp. 20, 73).

[46] A. A. Krasnov. "Molecular pumping properties of the LHC arc beam pipe and effective secondary electron emission from Cu surface with artificial roughness". In: *Vacuum* 73.2 (2004), pp. 195–199. ISSN: 0042-207X. DOI: 10/fckhwf (cit. on p. 20).

[47] E. Ebrahimibasabi and S. A. H. Feghhi. "Design and construction of a secondary electron suppressed Faraday Cup for measurement of beam current in an electrostatics proton accelerator". In: *International Journal of Mass Spectrometry* 386 (2015), pp. 1–5. ISSN: 1387-3806. DOI: 10/f7rzmx (cit. on p. 20).

[48] K. A. Avramides, I. G. Pagonakis, C. T. Iatrou, and J. L. Vomvoridis. "EURIDICE: A code-package for gyrotron interaction simulations and cavity design". In: *European Physical Journal Web of Conferences*. Vol. 32. Sept. 2012, p. 04016. DOI: 10/gdvqdt (cit. on pp. 21, 124).

[49] I. G. Pagonakis, S. Illy, Z. C. Ioannidis, T. Rzesnicki, K. A. Avramidis, et al. "Numerical investigation on spent beam deceleration schemes for depressed collector of a high-power gyrotron". In: *IEEE Transactions on Electron Devices* (2018), pp. 1–6. DOI: 10/gdvmsm (cit. on pp. 24, 174, 176).

[50] D. Bariou, C. Lievin, A. Beunas, and G. L. Cloarec. *Technical Report N5261T*. Tech. rep. Thales Electron Devices, Dec. 2003 (cit. on pp. 25, 44).

[51] T. H. Cormen, C. E. Leiserson, R. L. Rivest, and C. Stein. *Introduction to Algorithms*. 3rd ed. The MIT Press, Mar. 22, 2005. ISBN: 978-0-262-03384-8 (cit. on pp. 27, 157).

[52] A. Arsenjewa-Heil and O. Hell. "Eine neue Methode zur Erzeugung kurzer, ungedämpfter, elektromagnetischer Wellen großer Intensität". In: *Zeitschrift für Physik* 95.11 (Nov. 1935), pp. 752–762. ISSN: 0044-3328. DOI: 10/bth422 (cit. on p. 29).

[53] H. G. Kosmahl. "Modern multistage depressed collectors—A review". In: *Proc. IEEE* 70.11 (Nov. 1982), pp. 1325–1334. ISSN: 0018-9219. DOI: 10/bpc87s (cit. on pp. 29, 31–34).

[54] T. G. Mihran and W. Neugebauer. "Analytic Study of a Depressed Collector for Linear Beam Microwave Amplifiers". In: *NASA CR-72768* 7 (1970) (cit. on pp. 29–30, 32, 34).

[55] A. S. Gilmour. *Klystrons, Traveling Wave Tubes, Magnetrons, Crossed-Field Amplifiers, and Gyrotrons*. Artech House Microwave Library. Artech House, Incorporated, 2011. ISBN: 978-1-60807-185-2 (cit. on p. 29).

[56] H. G. Kosmahl. *A novel, axisymmetric, electrostatic collector for linear beam microwave tubes.* NASA technical note D-6093. Washington: National Aeronautics and Space Administration, 1971 (cit. on p. 29).

[57] J. R. Hechtel. "A novel electrostatic-focusing depressed collector for linear beam tubes". In: *IEEE Trans. Electron Devices* 24.1 (1977), pp. 45–52. ISSN: 0018-9383. DOI: 10/dxjw59 (cit. on p. 31).

[58] L. R. Elias and G. J. Ramian. "Status report of the UCSB FEL experimental program". In: *Free-electron generators of coherent radiation.* Ed. by C. A. Brau, S. F. Jacobs, and M. O. Scully. Vol. 453. Proc. SPIE. International Society for Optics and Photonics. Jan. 1984, pp. 137–146. DOI: 10/gdvmr4 (cit. on p. 31).

[59] I. Boscolo, G. Brautti, V. Stagno, and V. Variale. "A high efficiency multistage collector for a low quality electron beam recovery". In: *Nuclear Instruments and Methods in Physics Research Section A: Accelerators, Spectrometers, Detectors and Associated Equipment* 275.1 (1989), pp. 184–189. ISSN: 0168-9002. DOI: 10/bg37q4 (cit. on pp. 31–32).

[60] I. Boscolo, V. Stagno, V. Variale, and M. Bianconi. "Physics and simulations of a new design multistage collector inserted in an immersed-flow configuration electrostatic accelerator". In: *Nuclear Instruments and Methods in Physics Research Section A: Accelerators, Spectrometers, Detectors and Associated Equipment* 311.1–2 (1992), pp. 386–394. ISSN: 0168-9002. DOI: 10/fjs9np (cit. on pp. 31–32).

[61] T. Okoshi, E. Chiu, and S. Matsuki. "The tilted electric field soft-landing collector and its application to a traveling-wave tube". In: *IEEE Trans. Electron Devices* 19.1 (1972), pp. 104–110. ISSN: 0018-9383. DOI: 10/dv9rqh (cit. on pp. 32–33, 106, 117).

[62] D. A. Dunn, R. P. Borghi, and G. Wada. "A crossed-field multisegment depressed collector for beam-type tubes". In: *IRE Transactions on Electron Devices* 7.4 (Oct. 1960), pp. 262–267. ISSN: 0096-2430. DOI: 10/b63q8r (cit. on p. 33).

[63] W. Wien. "Untersuchungen über die electrische Entladung in verdünnten Gasen". In: *Annalen der Physik* 301.6 (1898), pp. 440–452 (cit. on p. 33).

[64] L. Zhang, W. He, A. W. Cross, A. D. R. Phelps, K. Ronald, et al. "Design of an energy recovery system for a gyrotron backward-wave oscillator". In: *IEEE Trans. Plasma Sci.* 37.3 (Mar. 2009), pp. 390–394. ISSN: 0093-3813. DOI: 10/fnrg7x (cit. on pp. 34–35, 44).

[65] L. Zhang, W. He, A. W. Cross, A. D. R. Phelps, K. Ronald, et al. "Numerical optimization of a multistage depressed collector with secondary electron emission for an X-band Gyro-BWO". In: *IEEE Trans. Plasma Sci.* 37.12 (Dec. 2009), pp. 2328–2334. ISSN: 0093-3813. DOI: 10/dpstt8 (cit. on pp. 34–35, 73).

[66] G. P. Scheitrum. *Depressed collector for sorting radial energy level of a gyrating electron beam.* US Patent 5,420,478. May 1995 (cit. on p. 35).

[67] E. M. Choi, C. D. Marchewka, I. Mastovsky, J. R. Sirigiri, M. A. Shapiro, et al. "Experimental results for a 1.5-MW, 110-GHz gyrotron oscillator with reduced mode competition". In: *Physics of Plasmas* 13.2 (Feb. 2006), p. 023103. DOI: 10/dzfz64 (cit. on pp. 36, 49).

[68] I. G. Pagonakis, J. P. Hogge, S. Alberti, K. A. Avramides, and J. L. Vomvoridis. "A new concept for the collection of an electron beam configured by an externally applied axial magnetic field". In: *IEEE Trans. Plasma Sci.* 36.2 (Apr. 2008), pp. 469–480. ISSN: 0093-3813. DOI: 10/b33nrk (cit. on pp. 37, 39–40, 79–80, 85, 121).

[69] J. G. Pagonakis and J. L. Vomvoridis. "The self-consistent 3D trajectory electrostatic code ARIADNE for gyrotron beam tunnel simulation". In: *Proc. Infrared and Millimeter Waves Conf.* Sept. 2004, pp. 657–658. DOI: 10/d4x52c (cit. on pp. 40, 44).

[70] J. Jelonnek, G. Aiello, S. Alberti, K. Avramidis, F. Braunmueller, et al. "Design considerations for future DEMO gyrotrons: A review on related gyrotron activities within EUROfusion". In: *Fusion Engineering and Design* 123 (Nov. 2017), pp. 241–246. DOI: 10/gct6h5 (cit. on p. 41).

[71] K. Sakamoto, M. Tsuneoka, A. Kasugai, T. Imai, T. Kariya, et al. "Major improvement of gyrotron efficiency with beam energy recovery". In: *Physical Review Letters* 73.26 (1994), pp. 3532–3535. ISSN: 0031-9007. DOI: 10/c3f7t5 (cit. on pp. 44, 132, 154).

[72] B. Piosczyk, C. T. Iatrou, G. Dammertz, and M. Thumm. "Single-stage depressed collectors for gyrotrons". In: *IEEE Trans. Plasma Sci.* 24.3 (June 1996), pp. 579–585. ISSN: 0093-3813. DOI: 10/dnn7g9 (cit. on pp. 44, 132, 172).

[73] C. Wu, I. G. Pagonakis, S. Illy, M. Thumm, G. Gantenbein, et al. "Preliminary studies on multistage depressed collectors for fusion gyrotrons". In: *10th German Microwave Conference*. Bochum, Germany, Mar. 2016. DOI: 10/gdvmrt (cit. on pp. 44, 62, 65–66).

[74] Y. N. Grigoryev, V. A. Vshivkov, and M. P. Fedoruk. *Numerical "Particle-in-Cell" Methods: Theory and Applications. Theory and Applications.* 2002. ISBN: 90-6764-368-8 (cit. on p. 45).

[75] J.-P. Hogge, F. Albajar, S. Alberti, P. Benin, T. Bonicelli, et al. "The European 2-MW, 170-GHz coaxial cavity gyrotron for ITER". In: *2007 Joint 32nd International Conference on Infrared and Millimeter Waves and the 15th International Conference on Terahertz Electronics*. IEEE, Sept. 2007. DOI: 10/gd3vdg (cit. on p. 46).

[76] G. Schmidt. *Physics of High Temperature Plasmas*. Academic Press, 1979. ISBN: 0-12-626660-3 (cit. on p. 65).

[77] M. V. Kartikeyan and B. Piosczyk. *Design of a Two-Stage Depressed Collector for a High Power Gyrotron*. Tech. rep. FE.130.0021.012/A. Forschungszentrum Karlsruhe, Sept. 1996 (cit. on pp. 66, 77).

[78] A. Singh, A. Valfells, M. J. Kolander, and V. L. Granatstein. "Improvements in depressed collector performance by modifications to electrode geometry vis-a-vis trajectories of backscattered electrons". In: *IEEE Transactions on Plasma Science* 32.3 (June 2004), pp. 1267–1276. ISSN: 0093-3813. DOI: 10/cvggjz (cit. on pp. 66, 73, 77).

[79] I. G. Pagonakis, F. Li, S. Illy, B. Piosczyk, S. Alberti, et al. "Study of the ITER stray magnetic field effect on the EU 170-GHz 2-MW coaxial cavity gyrotron". In: *IEEE Trans. Plasma Sci.* 40 (July 2012), pp. 1945–1956. DOI: 10/f39k9f (cit. on pp. 69–70).

[80] B. Ell, I. G. Pagonakis, G. Gantenbein, S. Illy, M. Thumm, et al. "Study of the influence of stray magnetic fields on the operation of the European gyrotron for ITER". In: *IEEE Trans. Electron Devices* 64 (Aug. 2017), pp. 3421–3428. DOI: 10/gdvmsf (cit. on pp. 69–70).

[81] C. C. Finlay, S. Maus, C. D. Beggan, T. N. Bondar, A. Chambodut, et al. "International geomagnetic reference field: the eleventh generation". In: *Geophysical Journal International* 183.3 (Oct. 2010), pp. 1216–1230. DOI: 10/bkmkk5 (cit. on p. 70).

[82] A. Samartsev, G. Dammertz, G. Gantenbein, J. Jelonnek, S. Illy, et al. "Influence of annular beam displacement on the performance of a high-power gyrotron". In: *IEEE Trans. Plasma Sci.* 41.4 (Apr. 2013), pp. 872–878. ISSN: 0093-3813. DOI: 10/f4vpns (cit. on p. 70).

[83] I. G. Pagonakis, C. Wu, S. Illy, and J. Jelonnek. "Multistage depressed collector conceptual design for thin magnetically confined electron beams". In: *Physics of Plasmas* 23.4 (2016), p. 043114. ISSN: 1070-664X. DOI: 10/f8mzh9 (cit. on pp. 80, 88, 91, 108, 111).

[84] I. Pagonakis, C. Wu, S. Illy, G. Gantenbein, and J. Jelonnek. "Multistage depressed collector design approach for gyrotron". In: *International Vacuum Electronics Conference (IVEC)*. IEEE, Apr. 2016. DOI: 10/gdvmrw (cit. on p. 80).

[85] C. Wu, I. Pagonakis, S. Illy, G. Gantenbein, M. Thumm, et al. "3D simulation of a realistic multistage depressed collector for high-power fusion gyrotrons". In: *International Vacuum Electronics Conference (IVEC)*. submitted. IEEE, Apr. 2016. DOI: 10/gdvmrv (cit. on pp. 80, 90).

[86] C. Wu, I. G. Pagonakis, G. Gantenbein, S. Illy, M. Thumm, et al. "Conceptual designs of E×B multistage depressed collectors for gyrotrons". In:

Physics of Plasmas 24.4 (Mar. 2017), p. 043102. DOI: 10/gdvqdv (cit. on pp. 80, 105, 110–111, 114–115, 118).

[87] C. Wu, I. G. Pagonakis, G. Gantenbein, S. Illy, M. Thumm, et al. "Design of E×B multistage depressed collector concepts for high-power fusion gyrotrons". In: *Proc. 42th Int. Conf. Infrared, Millimeter and Terahertz waves (IRMMW-THz)*. Aug. 2017, pp. 1–2. DOI: 10/gdvmr7 (cit. on p. 80).

[88] B. Ell, I. G. Pagonakis, C. Wu, and J. Jelonnek. "Conceptual design and simulation of a gyrotron multistage depressed collector based on E×B drift". In: *20th Joint Workshop on Electron Cyclotron Emission (ECE) and Electron Cyclotron Resonance Heating (ECRH)*. 2018 (cit. on pp. 80, 127, 145).

[89] O. I. Louksha and P. A. Trofimov. "A method of electron separation for multistep recuperation systems in gyrotrons". In: *Technical Physics Letters* 41.9 (2015), pp. 884–886. ISSN: 1063-7850. DOI: 10/gdvmrs (cit. on pp. 80, 171).

[90] O. I. Louksha and P. A. Trofimov. "A method for sorting of electrons in gyrotron multistage depressed collectors". In: *Proc. 40th Int. Conf. Infrared, Millimeter and Terahertz waves (IRMMW-THz)*. Aug. 2015, pp. 1–2. DOI: 10/gdvmsb (cit. on pp. 80, 171).

[91] O. I. Louksha and P. A. Trofimov. "A multistage depressed collector with azimuthal magnetic field for gyrotrons". In: *Proc. 41th Int. Conf. Infrared, Millimeter and Terahertz waves (IRMMW-THz)*. Sept. 2016, pp. 1–3. DOI: 10/gdvmr9 (cit. on pp. 80, 171).

[92] O. I. Louksha and P. A. Trofimov. "A multistage depressed collector system for gyrotrons". In: *Proc. Eighteenth Int. Vacuum Electronics Conf. (IVEC)*. Apr. 2017, pp. 1–2. DOI: 10/gdvmr8 (cit. on pp. 80, 86, 171).

[93] A. Clebsch. "Über eine allgemeine Transformation der hydrodynamischen Gleichungen." In: *Journal für die reine und angewandte Mathematik (Crelles Journal)* 1857.54 (Jan. 1857), pp. 293–312. ISSN: 0075-4102. DOI: 10/cbzj3s (cit. on p. 91).

[94] S. H. Lamb. *Hydrodynamics (Dover Books on Physics)*. Dover Publications, 1945. ISBN: 0-486-60256-7 (cit. on p. 91).

[95] M. Kruskal and R. Kulsrud. "Equilibrium of a magnetically confined plasma in a toroid". In: *Journal of Nuclear Energy* 7.3-4 (Sept. 1958), p. 297. ISSN: 0891-3919. DOI: 10/cwbx2d (cit. on p. 91).

[96] W. D. D'haeseleer, W. N. G. Hitchon, J. D. Callen, and J. L. Shohet. "The Clebsch-Type Coordinate Systems". In: *Flux Coordinates and Magnetic Field Structure: A Guide to a Fundamental Tool of Plasma Theory*. Springer Berlin Heidelberg, 1991. Chap. 5, pp. 100–115. ISBN: 978-3-642-75595-8 (cit. on p. 92).

[97] D. P. Stern. "Euler Potentials". In: *American Journal of Physics* 38.4 (Apr. 1970), pp. 494–501. DOI: 10/fg9qxr (cit. on p. 94).

[98] K. Watanabe. "Green's Functions for Laplace and Wave Equations". In: *Integral Transform Techniques for Green's Function*. Springer, Jan. 2014. ISBN: 978-3-319-00878-3 (cit. on p. 94).

[99] M. Schmid, S. Illy, G. Dammertz, V. Erckmann, and M. Thumm. "Transverse field collector sweep system for high power CW gyrotrons". In: *Fusion Engineering and Design* 82.5-14 (Oct. 2007), pp. 744–750. DOI: 10/bx4j6n (cit. on p. 102).

[100] R. W. Moir and R. F. Post. "Yin-yang minimum-$|B|$ magnetic-field coil". In: *Nuclear Fusion* 9.3 (1969), p. 243 (cit. on p. 104).

[101] A. Malygin, S. Illy, I. G. Pagonakis, B. Piosczyk, S. Kern, et al. "Design and 3D simulations of a 10-kW/28-GHz gyrotron with a segmented emitter based on controlled porosity-reservoir cathodes". In: *IEEE Trans. Plasma Sci.* 41.10 (Oct. 2013), pp. 2717–2723. ISSN: 0093-3813. DOI: 10/f5fqqd (cit. on p. 107).

[102] O. I. Louksha, B. Piosc, G. G. Sominski, M. K. Thumm, and D. B. Samsonov. "On potentials of gyrotron efficiency enhancement: measurements and simulations on a 4-mm gyrotron". In: *IEEE Trans. Plasma Sci.*

34.3 (June 2006), pp. 502–511. ISSN: 0093-3813. DOI: 10/c4prp9 (cit. on p. 108).

[103] V. E. Zapevalov, G. G. Denisov, V. A. Flyagin, A. S. Fix, A. N. Kuftin, et al. "Development of 170-GHz/1-MW Russian gyrotron for ITER". In: *Fusion Engineering and Design* 53.1-4 (Jan. 2001), pp. 377–385. ISSN: 0920-3796. DOI: 10/bv3z33 (cit. on p. 108).

[104] B. Piosczyk, G. Dammertz, O. Dumbrajs, O. Drumm, S. Illy, et al. "A 2-MW, 170-GHz coaxial cavity gyrotron". In: *IEEE Transactions on Plasma Science* 32.2 (Apr. 2004), pp. 413–417. DOI: 10 / b6nqq5 (cit. on p. 108).

[105] A. Bertinetti, K. A. Avramidis, F. Albajar, F. Cau, F. Cismondi, et al. "Multiphysics analysis of a 1-MW gyrotron cavity cooled by mini-channels". In: *Fusion Engineering and Design* 123 (Nov. 2017), pp. 313–316. DOI: 10/gct7fp (cit. on p. 121).

[106] V. N. Manuilov, M. V. Morozkin, O. I. Luksha, and M. Y. Glyavin. "Gyrotron collector systems: Types and capabilities". In: *Infrared Physics & Technology* 91 (2018), pp. 46–54. ISSN: 1350-4495. DOI: 10/gdvmsj (cit. on pp. 146, 171).

[107] M. Glyavin, V. Manuilov, and M. Morozkin. "Two-stage energy recovery system for DEMO gyrotron". In: *43rd International Conference on Infrared, Millimeter, and Terahertz Waves (IRMMW-THz)*. IEEE, Sept. 2018. DOI: 10/czxr (cit. on p. 146).

[108] H. G. Kosmahl. "How to quickly predict the overall TWT and the multistage depressed collector efficiency". In: *IEEE Trans. Electron Devices* 27.3 (Mar. 1980), pp. 526–529. DOI: 10/bqx6g7 (cit. on p. 157).

[109] M. Iri, S. Moriguti, and Y. Takasawa. "On a certain quadrature formula". In: *Journal of Computational and Applied Mathematics* 17.1-2 (Feb. 1987), pp. 3–20. DOI: 10/b9q8fh (cit. on p. 162).

List of Figures

1.1 Schematic of a general EC system for ITER [22] 5

1.2 Schematic of a gyrotron 7

1.3 Cross section of the annular electron beam in a small-orbit gyrotron 8

1.4 Relation between power and efficiencies in a gyrotron 10

1.5 Relation of η_{int}, η_{col} and η_{total} for $\eta_{\text{RF}} = 90\,\%$ 11

1.6 Schematic of the electron beam cross section showing the magnetic fluxes in eqs. (1.21) and (1.25) 16

1.7 Model of secondary electron energy distribution by a perpendicular hit of a 1 keV primary electron 17

1.8 SEY curve for perpendicularly impacted electrons 18

1.9 Spent electron beam spectra of the 140 GHz gyrotron for W7-X and the 170 GHz 1 MW gyrotron for ITER 21

1.10 Deceleration of the electron beam in an ideal SDC 22

1.11 The reference SDC for the 170 GHz 1 MW gyrotron 23

1.12 Deceleration scheme for an ideal two-stage depressed collector . . . 26

1.13 Relation between efficiencies and number of stages 28

1.14 Classification chart of TWT MDCs, reconstructed from [54] 30

1.15 Axisymmetric dispersive lens collector [53] (for TWT) 31

1.16 Trajectories in the asymmetric MDC scheme for FEL [59], the repulsion of space charges is ignored 32

1.17 Schematic of the tilted electric field collector [61] 33

1.18 The first sub-type of axisymmetric MDCs for a 110 GHz, 1 MW gyrotron [10] 36

1.19 The second sub-type of axisymmetric MDCs using a so-called "magnetic separator" [12] 36
1.20 Scheme of an $\boldsymbol{E}\times\boldsymbol{B}$ MDC using $E_\perp$ 38
1.21 The original idea of gyrotron $\boldsymbol{E}\times\boldsymbol{B}$ MDC with an infinite number of stages [68] 40
2.1 Principle of the advanced axisymmetric MDC based on the spatial modulation of electron trajectories 51
2.2 Projection of the cyclotron motion in a homogeneous magnetic field 53
2.3 Relative values of the initial velocity components in a realistic axisymmetric MDC design 56
2.4 Visualization of the r_{gc} term in eq. (2.14) 57
2.5 Larmor radii after a non-adiabatic transition 60
2.6 Cyclotron longitudinal periods after a non-adiabatic transition . . . 60
2.7 Axisymmetric MDC design proposal 64
2.8 Effective potential 67
2.9 Examination of the sorting quality using mono-energetic electron beams for the axisymmetric MDC design 68
2.10 Sensitivity of the perturbation radial magnetic field 71
2.11 Sensitivity of the beam displacement in the cavity 71
2.12 Sensitivity of electron beam current 72
2.13 Trajectories of secondary electrons 74
2.14 Power load density of the axisymmetric MDC proposal 75
3.1 Category of MDC schemes based on the $\boldsymbol{E}\times\boldsymbol{B}$ drift 80
3.2 Scheme of an MDC for a sheet electron beam 81
3.3 Increasing the angle ϑ between the electric field and the magnetic field will also increase the collector length 84
3.4 Cross sections of two possibilities to produce radial $\boldsymbol{E}\times\boldsymbol{B}$ drifts . . . 85

3.5 Relation of the drift distance D and B_θ in the $\boldsymbol{E}\times\boldsymbol{B}$ drift concept using a transverse magnetic field with a fixed B_z 87
3.6 Example for the transformation of the magnetic field lines from an annular cross-section profile to a profile of two sheets . . . 88
3.7 CST simulation to verify the concept of sheet beam transformation [85] 90
3.8 Schematic of the iso-α and iso-β surfaces for the modeling of the beam-shape transformation 93
3.9 Optimization of the sheet-beam transformation by manipulating the shapes of the iso-α surfaces 96
3.10 Upgraded transformation coils to avoid the unwanted non-adiabatic magnetic field produced by the transverse current 99
3.11 Upgraded electrodes for the sheet beam collector 100
3.12 Triple boxes make the collector shorter, however, more complex . . 101
3.13 Typical thermal load at the last three stages of a sheet-beam MDC . 103
3.14 Cross section of an elongated elliptic electron beam 104
3.15 Transformation from annular electron beam to elongated elliptic beam using ideal coils [86] 105
3.16 Elongated elliptic beam transformation using more reasonable coils [86] 105
3.17 A side view of the electric field, magnetic field and the electrodes for an elliptic-beam MDC 107
3.18 Breaking an elliptic electron beam into two curved sheet beams . . . 109
3.19 Transverse projections of the (initial) arc electron beams and the electrodes [86] 110
3.20 Estimated total length of a three-stage arc-beam collector 114
3.21 Design of a three-stage arc-beam depressed collector with the simulated electron trajectories [86] 115
3.22 Suppressing secondary electrons in the arc-beam MDC [86] 118
3.23 Thermal load at the baffles 120

3.24 Idea of an MDC with helical electrodes 122
3.25 The choice of design parameters based on eq. (3.34) [14] 123
3.26 Beam cross section (taken at position ① of fig. 3.24) in a steady-state trajectory simulation [14] 125
3.27 Examination with mono-energetic electron beam 126
3.28 Sensitivity of efficiency for ±20 % beam current [14] 128
3.29 Sensitivities of efficiency for anisotropic misalignment and perturbation $B_\perp$ [14] 129
3.30 Convergence of the RF power in the ARIADNE full-gyrotron simulation with 1.2 A of 45 A back-stream current 131
3.31 Power load density at the first stage 134
3.32 Power load density at the second stage 134
3.33 Azimuthal drift produced by a coaxially inserted electrode [15] . . . 136
3.34 The helical structure covers more than 2π azimuthal range [15] . . . 139
3.35 Disk at the entrance catching the reflections [15] 140
3.36 Combination of figs. 3.34 and 3.35 [15] 141
3.37 Examination with mono-energetic electron beam 142
3.38 Azimuthal repetition of helices [15] 143
3.39 Mechanical design proposal of a two-stage depressed collector . . . 147
3.40 A module of the first stage 147
3.41 Assembling the components of the mechanical design proposal . . . 149
A.1 Sketch of a current loop 159
A.2 Current segment I induces a field at (x_0, y_0, z_0) 165
B.1 Proposal of a coil scheme for an $\boldsymbol{E}\times\boldsymbol{B}$ MDC using B_θ [106] 171
B.2 Axisymmetric MDC based on the sudden change of electric field . . 173
B.3 Geometry of the MDC with inward drift 174
B.4 Top view of the MDC with inward drift 175

Index

A

action 13
Ampère's law 171
angular momentum 14
Ariadne 40, 44

B

Biot-Savart integral 160
Busch's theorem 15

C

canonical angular momentum . 15
Cauchy-Schwarz inequality ... 55
cavity 6
charge density 45
collector efficiency 9
Coulomb force 3
Coulomb gauge 92
cyclotron frequency 5

D

demagnetization 35
dispersive lens 29
displacement of electron beam . 70
drift distance 80
dynamic programming ... 27, 157

E

effective potential 65
eigen frequency 6
elastically reflected electron ... 17
electric potential 45
electron beam 8
electron cyclotron frequency ... 4
equivalent current loop 112
Esray 44
Euridice 21

F

far field of a current loop 112
Faraday's law 39, 85
ferromagnet 37
free electron laser 31
full-gyrotron simulation 131
fusion break-even 3

G

Green's function 94
guiding center 8

H

Hamiltonian of a test particle .. 66

I

inelastically reflected electron . 17
inertial confinement 3
interaction 1, 6

L

Lagrangian of a charged particle 15
Larmor radius 5, 13
linear electron beam 29
Lorentz factor 4
Lorentz force 12
low-perveance beam 67

M

magnetic confinement 3
magnetic field flux 106
magnetic moment 14
magnetic separator 37
misalignment of electron beam 70

P

particle in cell 45
permittivity 45
perturbation magnetic field . . . 70
pitch factor 6
Poisson equation 45, 94
pole shoe 35
power load 25, 41

Q

quadrupole 95

R

reference SDC 25
RF efficiency 9
RF power 131

S

secondary electron 16, 72
  yield 17
sensitivity 127
small-orbit electron beam 8
spent electron beam 8, 21
  energy spectrum 21
stellarator 3
sweeping 25

T

tilted electric field 32
tokamak 3
trajectory tracking 45
transverse velocity 6
true secondary electron 17

U

universal beam-spread curve . . 29

V

vacuum envelope 146
value function 158

W

Wien filter 33

KARLSRUHER FORSCHUNGSBERICHTE AUS DEM INSTITUT FÜR HOCHLEISTUNGSIMPULS- UND MIKROWELLENTECHNIK (ISSN 2192-2764)

Band 1 **MATTHIAS BERINGER**
Design Studies towards a 4 MW 170 GHz Coaxial-Cavity Gyrotron. 2011
ISBN 978-3-86644-663-2

Band 2 **JENS FLAMM**
Diffraction and Scattering in Launchers of Quasi-Optical Mode Converters for Gyrotrons. 2012
ISBN 978-3-86644-822-3

Band 3 **MATTIA DEL GIACCO**
Investigation of Fretting Wear of Cladding Materials in Liquid Lead. 2013
ISBN 978-3-86644-960-2

Band 4 **AMITAVO ROY CHOUDHURY**
Investigations of After Cavity Interaction in Gyrotrons Including the Effect of Non-uniform Magnetic Field. 2013
ISBN 978-3-7315-0129-9

Band 5 **MICHAEL BETZ**
The CERN Resonant WISP Search (CROWS). 2014
ISBN 978-3-7315-0199-2

Band 6 **ANDREAS SCHLAICH**
Time-dependent spectrum analysis of high power gyrotrons. 2015
ISBN 978-3-7315-0375-0

Band 7 **DHIDIK PRASTIYANTO**
Temperature- and Time-Dependent Dielectric Measurements and Modelling on Curing of Polymer Composites. 2016
ISBN 978-3-7315-0424-5

Band 8 **YIMING SUN**
Adaptive and Intelligent Temperature Control of Microwave Heating Systems with Multiple Sources. 2016
ISBN 978-3-7315-0467-2

Die Bände sind unter www.ksp.kit.edu als PDF frei verfügbar oder als Druckausgabe bestellbar.

Band 9 **JIANGHUA ZHANG**
Influence of Emitter surface roughness and Emission inhomogeneity on Efficiency and stability of high power Fusion gyrotrons. 2016
ISBN 978-3-7315-0578-5

Band 10 **ANTON MALYGIN**
Design and Experimental Investigation of a Second Harmonic 20 kW Class 28 GHz Gyrotron for Evaluation of New Emitter Technologies. 2016
ISBN 978-3-7315-0584-6

Band 11 **JOACHIM FRANCK**
Systematic Study of Key Components for a Coaxial-Cavity Gyrotron for DEMO. 2017
ISBN 978-3-7315-0652-2

Band 12 **PARTH CHANDULAL KALARIA**
Feasibility and Operational Limits for a 236 GHz Hollow-Cavity Gyrotron for DEMO. 2017
ISBN 978-3-7315-0717-8

Band 13 **CHUANREN WU**
Conceptual Studies of Multistage Depressed Collectors for Gyrotrons. 2019
ISBN 978-3-7315-0934-9

Die Bände sind unter www.ksp.kit.edu als PDF frei verfügbar oder als Druckausgabe bestellbar.